T0215668

Mobile and Wireless Communications with Practical Use-Case Scenarios

The growing popularity of advanced multimedia-rich applications along with the increasing affordability of high-end smart mobile devices has led to a massive growth in mobile data traffic that puts significant pressure on the underlying network technology. However, no single network technology will be equipped to deal with this explosion of mobile data traffic. While wireless technologies had a spectacular evolution over the past years, the present trend is to adopt a global heterogeneous network of shared standards that enables the provisioning of quality of service and quality of experience to the end-user. To this end, enabling technologies like machine learning, Internet of Things and digital twins are seen as promising solutions for next generation networks that will enable an intelligent adaptive interconnected environment with support for prediction and decision making so that the heterogeneous applications and users' requirements can be highly satisfied.

The aim of this textbook is to provide the readers with a comprehensive technical foundation of the mobile communication systems and wireless network design, and operations and applications of various radio access technologies. Additionally, it also introduces the reader to the latest advancements in technologies in terms of Internet of Things ecosystems, machine learning and digital twins for IoT-enabled intelligent environments. Furthermore, this textbook also includes practical use-case scenarios using Altair WinProp Software as well as Python, TensorFlow and Jupyter as support for practice-based laboratory sessions.

Dr. Ramona Trestian is a Senior Lecturer with the Design Engineering and Mathematics Department, Faculty of Science and Technology, Middlesex University, London, UK. She was previously an IBM-IRCSET Exascale Postdoctoral Researcher with the Performance Engineering Laboratory (PEL) at Dublin City University (DCU), Ireland. She was awarded the PhD from Dublin City University in March 2012 and the B.Eng. in Telecommunications from Technical University of Cluj-Napoca, Romania, 2007. She has published in prestigious international conferences and journals and has five edited books. Her research interests include mobile and wireless communications, quality of experience, multimedia systems, Industry 4.0 and digital twin modelling. She is an Associate Editor of the IEEE Communications Surveys and Tutorials.

Mobile and Wireless Communications with Practical Use-Case Scenarios

Ramona Trestian

CRC Press
Taylor & Francis Group
Boca Raton London New York

CRC Press is an imprint of the
Taylor & Francis Group, an **informa** business

First edition published 2023
by CRC Press
6000 Broken Sound Parkway NW, Suite 300, Boca Raton, FL 33487-2742

and by CRC Press
4 Park Square, Milton Park, Abingdon, Oxon, OX14 4RN

© 2023 Ramona Trestian

CRC Press is an imprint of Taylor & Francis Group, LLC

Library of Congress Cataloging-in-Publication Data

Names: Trestian, Ramona, 1983- author.
Title: Mobile and wireless communications with practical use case scenarios
  / Ramona Trestian.
Description: Boca Raton : CRC Press, 2023 | Includes bibliographical
  references and index.
Identifiers: LCCN 2022028937 (print) | LCCN 2022028938 (ebook) | ISBN
  9781032119014 (hardback) | ISBN 9781032119021 (paperback) | ISBN
  9781003222095 (ebook)
Subjects: LCSH: Wireless communication systems--Textbooks | System
  design--Textbooks | Use cases (Systems engineering)--Textbooks
Classification: LCC TK5103.2 .T695 2023 (print) | LCC TK5103.2 (ebook) |
  DDC 621.382--dc23/eng/20221017
LC record available at https://lccn.loc.gov/2022028937
LC ebook record available at https://lccn.loc.gov/2022028938

ISBN: 978-1-032-11901-4 (hbk)
ISBN: 978-1-032-11902-1 (pbk)
ISBN: 978-1-003-22209-5 (ebk)

DOI: 10.1201/9781003222095

Typeset in CMR10
by KnowledgeWorks Global Ltd.

Publisher's note: This book has been prepared from camera-ready copy provided by the author.

To all the women in STEM!
To my little sunshine, Noah Anthony,
my dear husband, Kumar,
and my lovely parents, Maria and Vasile.

Contents

III   Paradigms of Intelligent Networked Systems   235

Preface

The advances in technologies along with the growing popularity of video-based applications and the affordability of the latest smart mobile devices led to a massive growth in mobile data traffic that puts significant pressure on the underlying network technology. However, no single network technology will be equipped to deal with this explosion of data traffic. While wireless technologies had a spectacular evolution over the past years, the present trend is to adopt a global heterogeneous network of shared standards that enable the provisioning of quality of service and quality of experience to the end-user. To this end, enabling technologies like machine learning, Internet of Things and digital twins are seen as promising solutions for next generation wireless networks that will enable an intelligent adaptive interconnected environment with support for prediction and decision making so that the heterogeneous applications and users' requirements can be highly satisfied.

The aim of this textbook is to provide the readers with a comprehensive technical foundation of the mobile communication systems and wireless network design and operations, and applications of various radio access technologies. Additionally, it also introduces the reader to the latest advancements in technologies in terms of Internet of Things ecosystems, machine learning and digital twins for IoT-enabled intelligent environments. The novelty of this textbook is that it will include practical use-case scenarios using Altair Win-Prop Software as well as Python, TensorFlow and Jupyter to support the chapters for practice-based laboratory sessions.

The book is structured in three parts: (1) the first part will cover all the fundamental aspects of signals, analogue and digital communication systems, covering topics related to antennas and radio propagation; (2) the second part is concerned with the evolution of mobile and wireless systems, starting with the concept of cellular systems and applications, covering the cellular systems evolution from 1G to 5G and beyond, as well as wireless evolution and satellite communications; (3) finally, the third part introduces the paradigms of intelligence-based networked systems and in particular Internet of Things (IoT), machine learning and more recently digital twins that enable intelligent habitation, intelligent building control, smart vehicle technology, industrial processes and so forth.

List of Figures

List of Tables

Part I

Fundamental Aspects of Signals, Analogue and Digital Communication Systems

Chapter 1

The Wireless Vision

"I designed the executive program for handling situations when there are too many calls, to keep it operating efficiently without hanging up on itself. Basically it was designed to keep the machine from throwing up its hands and going berserk."

Dr. Erna Schneider Hoover
Mathematician and Inventor of the
computerized telephone switching method at Bell Laboratories

1.1 Introduction to Wireless Communication – Evolution and History

Mobile and wireless technologies had a spectacular evolution over the past decade, and the present trend is to move towards a highly intelligent global digital ecosystem of shared standards that could enable unlimited wireless connectivity and seamless Quality of Experience (QoE) for the mobile users. People have always been interested in different ways to convey important information and transmit messages from one place to another quickly and efficiently. This aspect drove the human innovation starting with the prehistoric human that made use of smoke signals to today's use of high-end mobile devices. Some of the key milestones that demonstrate the never-ending push for progress in the science and technology of communication are highlighted below [2; 3]:

During the **prehistoric era**, smoke signals were used as the first attempt for long-distance communication. Important information with pre-decided meanings like *danger* or *victory* was encoded into the smoke signal and transmitted over a limited geographic area. For example, smoke signals were used by soldiers in Ancient China to communicate from tower to tower along the Great Wall about the impeding enemy attack.

DOI: 10.1201/9781003222095-1

In **490 BC** during the first Persian invasion of Greece, the Battle of Marathon took place. Pheidippides, who was a Greek soldier and messenger, ran over 25 miles from Marathon to Athens to inform the Athenians about the victory over the Persians. The story relates that after delivering the message, Pheidippides dramatically dropped dead. This tale is a proof that the communication signal was really bad back then.

During **World War I**, carrier pigeons were used to transport important information, especially in hard-to-reach areas where radio use was impossible. A wartime story relates that the messenger pigeon, named Cher Ami, once saved around 200 men from friendly fire during the actions of the so-called "Lost Battalion".

In **1876** Alexander Graham Bell made the first successful telephone transmission to his assistant, Thomas A. Watson saying the now-famous phrase: "Mr. Watson, come here. I want to see you!". To everyone's delight Mr. Watson declared that he understood each word clearly, which is still not the case more than a century later when people would be "What? Say that again?".

In **1895** Guglielmo Marconi put Nikola Tesla's wireless discoveries into practice by actually sending and receiving the first radio signal in Italy. This marks the start of the commercial radio.

In **1906** Reginald Aubrey Fessenden was the first to demonstrate the two-way radio-telegraphic communication across the Atlantic Ocean. It was the first wireless radio telephone.

In **1908** Nathan B. Stubblefield was awarded the first US patent on a wireless phone. However, his device employed conduction and inductive fields and not the standard radio transmission of electromagnetic radiation as we know today.

In **1926** radio telephone connection was introduced for the first-class passengers on trains in Germany, between Hamburg and Berlin.

1939 marks the start of the World War II where the technology played a significant role, and Germans started using radio phones on a larger scale.

In **1947** Bell Labs proposed the use of hexagonal cells for mobile phones at the theoretical level, with the three-sided antenna as we know today.

In **1956** Ericsson launched the first fully automatic mobile phone in Sweden. The handset was referred to as Mobiltelefonisystem A or MTA, and had a weight of 90 pounds. However, 10 years later, in **1965** Ericsson launches MTB. This time the handset weighs 20 pounds only, due to the use of transistors.

In **1970** the "call handoff" system is invented by Bell Laboratories. This allows the mobile phones to seamlessly move through the cellular network, moving from a cell area to another cell are without connectivity loss during an active conversation.

1971 marks the start of the "zero-generation (0G)" in Finland through ARP (Autoradiopuhelin or the "car radio phone") system which was the first successful commercial mobile phone network. However, there was no seamless support for users to move from cell to cell.

In **1973** Dr. Martin Cooper from Motorola used the first Motorola DynaTAC prototype to call Joel Engel, the head of research at AT&T Bell Labs while walking in New York City. This marks the beginning of the "first-generation" (1G) networks. However, roaming was not possible, there was no security and the downlink speed reached up to 2.4 kbps.
In **1978** Bell Labs launched the first trial for commercial cellular networks in Chicago.

In **1982** the first mobile phone was introduced by Nokia, the analog Mobira Senator. Additionally, the Federal Communications Commission (FCC) approved the analog-based Advanced Mobile Phone Service (AMPS) and assigned frequencies in the 824-894 MHz band.

1983 marks the year when the first call ever was made with a commercially available mobile phone such as Motorola DynaTAC. The weight of the Motorola DynaTac was around 2 pounds, requiring 10 hours of charging time for 30 minutes of talk time.

In **1988** Mark Weiser, the chief scientist at Xerox PARC, expressed his vision on *ubiquitous computing* [4] stating that there will be "...hundreds of wireless computing devices per person per office, of all scales [...] This is different from Personal Digital Assistant (PDA)'s, dynabooks, or information at your fingertips. It is invisible, everywhere computing that does not live on a personal device of any sort but is in the woodwork everywhere. [...] its highest ideal is to make a computer exciting, so wonderful, so interesting, that we never want to be without it".

In **1990** FCC approved the digital AMPS, which marks the beginning of the end for the analog networks.

In **1991** Nokia hardware was used to make the first commercial Global System for Mobile Communications (GSM) call in the world. This marks the beginning of "second-generation" (2G) networks and digital innovation. 2G provided encrypted calls and downlink speed of up to 0.2 Mbps.

1993 marks the appearance of the Short Message Service (SMS) supported by the Nokia handset. (e.g., txt msgng apprs 4 1st time LOL)

In **1996** Motorola launched its Motorola StarTAC handset. This is consider to be the first clam-shell mobile phone that gained widespread consumer adoption.

In **2001** NTT DoCoMo deployed the "third-generation" (3G) network to the public in Japan. 3G enabled international roaming and downlink speed of up to 2 Mbps. 3G also offered support for more advanced applications like video streaming, video conferences and live video chat. In **2002**, FCC decided to shut down the analog network.

In **2002** Kevin Ashton, the executive director of the Auto-ID Centre at Massachusetts Institute of Technology (MIT), coined the Internet of Things (IoT) term: "We need an internet for things, a standardized way for computers to understand the real world" [5].

In **2005** the International Telecommunication Union (ITU) published the first report on the topic around IoT stating that: "always on communications, in which new ubiquitous technologies (such as radio frequency identification and sensors) promise a world of networked and interconnected devices (e.g. fridge, television, vehicle, garage door, etc.) that provide relevant content and information whatever the location of the user heralding the dawn of a new era, one in which the internet (of data and people) acquires a new dimension to become an Internet of Things." [6].

In **2007** Apple launched its first iPhone. In only few years, it dominated the smartphone marketplace.

In **2009** the "fourth-generation" (4G) network was deployed for commercial use in Norway. 4G enabled the support for High Definition (HD) videos, online gaming, fast mobile web access, high-quality video streaming/video calls and a minimum speed availability of 12.5 Mbps.

In **2013** Cisco coined the Internet of Everything (IoE) term: "The IoE brings together people, processes, data, and things to make networked connections more relevant and valuable than ever before turning information into actions that create new capabilities, richer experiences, and unprecedented economic opportunity for businesses, individuals, and countries" [7].

In **2019** the "fifth-generation" (5G) was first rolled out by South Korea. 5G offers reduced latency, increased data speeds and new network capabilities that could lead to the next digital revolution.

A roadmap of the wireless and cellular networks evolution is illustrated in Figure 1.1.

The COVID-19 global pandemic has significantly disrupted different sectors gaining its name as the *great accelerator* of the digital transformation. This is because it speeds-up the adoption of digital technologies by several years across different industry sectors including education [8; 9] and healthcare [10; 11]. Moreover, the *great accelerator* pushed the development of Industry 4.0 that together with the rapid digitization across different industries are paving the way for the fifth industrial revolution. The vision of Industry 5.0 is to create an ecosystem of coexistence between people, processes and machines enabled by the autonomous manufacturing progress with human intelligence in and on the loop. Industry 5.0 revolution could also see the elevation of the level of virtual interactions with the physical environment, enabled by the continuous industrial adoption of the Digital Twin (DT) technology [12].

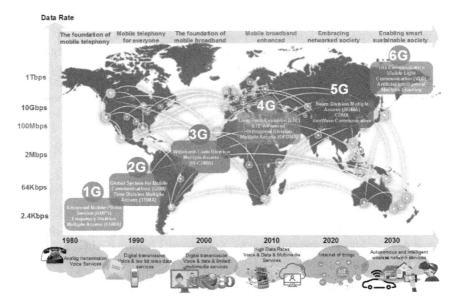

FIGURE 1.1
Wireless Networks Roadmap Outlook

The DT concept goes beyond the traditional computer-based simulations and analysis and it represents a two-way communication bridge between the physical world and the digital world [13]. The physical object exists in symbiotic relationship with its digital counterpart, being connected through real-time data communications and information transfer. The existence and the fast adoption of DTs is powered by the advances in key enabling technologies, including IoT, Artifical Intelligence (AI), Augmented Reality (AR), Virtual Reality (VR), Big Data Analytics, Multi-access Edge Computing (MEC), next-generation networks, edge/cloud/fog computing, etc. The integration of these enabling technologies could help at extending the capabilities of both machines and people, in order to optimize efficiency and operations within various industries.

The advances of 5G networks were expected to revolutionize the way we communicate, perceive and compute data by enabling a ubiquitous and pervasive paradigm ecosystem. Among the major objectives of 5G, one is to enable the QoS provisioning among three service classes, such as enhanced Mobile Broadband (eMBB), Ultra-Reliable and Low Latency Communications (URLLC) and massive Machine Type Communications (mMTC). However, even though 5G networks are currently being deployed around the world, research efforts are put by many industry pioneers and technology leaders towards defining the next-generation 6G networks. The idea behind 6G is to enhance even further all the applications and vertical use cases of the 5G network by bringing the intelligence at the edge of the network [14]. Additionally, 6G envisions interactions between three worlds: the human world (e.g., senses, bodies, intelligence, etc.), the digital world (information, communication, computing, etc.) and the physical world (objects, organisms, processes, etc.) [15]. Consequently, what lies beyond 5G would be an intelligent interconnected system of DTs that enables the creation of a real-time digital world. Thus, 6G will be represented by connected and augmented intelligence that will change the way the data is created, processed and consumed [13]. Furthermore, 6G envisions new service classes such as Further enhanced Mobile Broadband (FeMBB), enhanced Ultra-Reliable and Low Latency Communication (eURLLC) and ultra-massive Machine Type Communication (umMTC) that will require speeds of up to 1000 times faster than 5G [14] as well as latency less than 1ms to enable support for mission critical communications and Industrial Internet of Things (IIoT) applications [16].

1.2 Applications and Technical Challenges

The rapidly evolving Information and Communications Technology (ICT) industry currently faces a stealth, but very powerful service-oriented revolution [17]. This is firstly due to the smart mobile computing devices that have become increasingly powerful and affordable which led to a significant growth in their number, as well as their users number and their technical capabilities in terms of processing power, improved display and graphics and communication. Additionally, there is a wide range of multimedia-rich applications that are increasingly popular, such as YouTube, Twitter, Instagram, Facebook, that are known to be power and bandwidth-hungry applications that put pressure on both content processing as well as the underlying delivery infrastructure [18]. An increasing trend to alleviate the challenges of trying to put up with the increasing computational power required by future applications, is to move towards cloud computing for processing and communication needs. This could offer both robustness and flexibility. Moreover, the wide range of technologies, available nowadays, enable the mobile users to be *Always Best Connected* to the Internet and have access to a variety of services from anywhere and at any time, while on the move or stationary. Another important trend that contributes to this service-oriented revolution is the increasing popularity of IoT adoption that seeks the inter-communication between a wide range of *things* including machines, appliances, sensors and vehicles. In this context, there is a global vision in finding sustainable, energy-efficient solutions for supporting the new rich media diverse services while maintaining the best quality level, given service requirements, network technology availability, network delivery conditions, end-user device characteristics as well as user profiles.

With the advent of 6G networks , there is an increasing demand to leverage Machine Learning (ML) capabilities and develop new and innovative smart applications that could enhance accuracy, user experience, efficiency and capabilities.

According to Cisco [19], the primary contributors to the global mobile traffic growth are the different mix of wireless devices, including smartphones, Machine to Machine (M2M), tablets, Personal Computer (PC)s, etc. It is estimated that by 2023 there will be 4.4 billion M2M connections represented by Global Positioning System (GPS) in cars, asset tracking systems in shipping and manufacturing sectors, as well as medical applications for patient records and health status. The advances in IoT, VR, AR, ML and AI pave the way for future applications, such as: self-driving vehicle diagnostics, Ultra High Definition (UHD) VR, cloud gaming, UHD streaming, etc. However, these multimedia-rich applications will require strict QoS requirements that need to be supported on a heterogeneity of hardware platforms associated with the content access devices and dynamic wireless network connectivity which might hamper their potential.

The literature provides a wide range of solutions that aim to overcome these challenges and improve QoS, including energy-efficient cluster-oriented multimedia delivery [20], quality-aware adaptive multimedia [21; 22], radio resource scheduling over dense networks [23], Reinforcement Learning (RL)-based scheduling [24] or load balancing solutions [25], etc.

1.2.1 Multimedia-Based Network Delivery Solutions

Network delivery of multimedia-rich applications refers to the process of delivering multimedia applications like movies, video clips, live presentations, etc. over a network either in real time or non-real time. When dealing with the network delivery of multimedia-rich applications, we can identify two distinct methods [18; 26; 27]:

- *downloading* – simplest form of multimedia delivery divided into two categories [28]: (1) *traditional download* – where the user downloads the video file from a web server on to their mobile device. The user can locally watch the video after the video file has been fully downloaded. This method does not assume any real-time performance expectations. (2) *progressive download* – the video content is downloaded from a web server on their mobile device. The user can locally watch the multimedia content as it is received by the mobile device and does not have to wait for the download to complete.

- *streaming* – makes use of a specialized multimedia streaming server which delivers, on request, the multimedia content required by the user. The video file is not downloaded on the user's mobile device and the user can play the multimedia content as it is delivered. Two categories can be identified as illustrated in Figure 1.2 [28]: (1) *traditional streaming* – the web

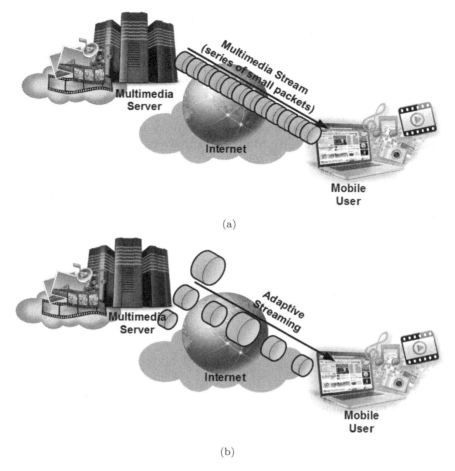

(a)

(b)

FIGURE 1.2
Multimedia Streaming Solutions: (a) Traditional Streaming and (b) Adaptive Streaming

server continuously monitors the user's state (e.g., Play, Seek or Pause) over the connection duration, and sends enough data packets to fill the user's buffer. (2) *adaptive streaming* – represents a hybrid delivery method that combines the streaming and progressive download methods. The multimedia content is encoded at different encoding rates (quality levels) and divided into small chunks that are stored at the server. A decision mechanism will switch between different quality levels to be delivered to the user based on their connection quality.

In order to overcome the impact of poor network connectivity or dynamic wireless environments on the users' perceived quality of the multimedia services, adaptive multimedia solutions have been widely explored in the recent literature: [29–33]. The main idea of these approaches is to adjust the amount of data transmitted over the underlying network in order to improve the end-user QoE. The adaptation decision is done either at the client side or at the server side, taking into consideration diverse objectives like video perceived quality, energy consumption, load balancing, cost, device type, etc. [34–43]

An example of an adaptive multimedia delivery solution, referred to as Signal Strength-based Adaptive Multimedia Delivery Mechanism (SAMMy) [26; 44], is represented in Figure 1.3.

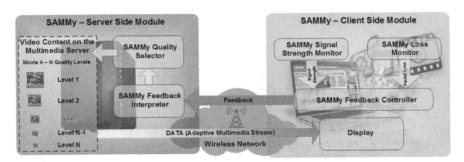

FIGURE 1.3
SAMMy Architecture

SAMMy is a distributed solution and consists of a server-side module which streams real-time multimedia content over wireless networks, and a client-side module which attaches to the multimedia client application, receives and displays the multimedia stream content.

SAMMy Server-side module is composed of three sub-modules: the Video Content, SAMMy Quality Selector, and SAMMy Feedback Interpreter. The video content stored on the server is encoded at different quality levels (e.g., different frame rate, frame size, bit rate, etc.). Consequently for a Movie A, the multimedia server can store a number of N Quality Levels (with Level 1 – the highest quality level to Level N – the lowest quality level). These quality levels correspond to different amounts of data to be delivered. SAMMy Feedback Interpreter receives feedback information, containing statistical data regarding the packet loss from the mobile device. Based on this received feedback information, SAMMy Feedback Interpreter will trigger the SAMMy Quality Selector which selects the most suitable quality level and consequently adjusts the multimedia delivery rate that is sent back to the mobile device. A more detailed description of the principles behind SAMMy is provided in the next section.

SAMMy Client-side module comprises of three sub-modules: SAMMy Signal Strength Monitor, SAMMy Loss Monitor and SAMMy Feedback Controller. SAMMy Signal Strength Monitor is responsible for monitoring the received signal strength (link quality) of the mobile device and it has two operation modes: (1) instant reading mode which triggers the module to measure the instantaneous received signal strength of the mobile device and (2) the prediction-based mode which predicts the received signal strength for a future location of the mobile user. In order to predict the received signal strength, information about the user current location within the wireless network (relative to the AP) is needed.

1.2.2 360^0 Video Streaming Applications

The current advancements in mobile and wireless technologies, enabled very low latency, higher data rates and more capacity being able to accommodate new generations of video streaming technologies, such as 360^0 or omnidirectional videos [45]. Figure 1.4 illustrates a 360^0 video viewing experience, where the user is wearing a Head-Mounted Display (HMD) device to be able to experience the 360^0 video streaming.

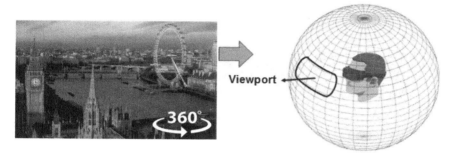

FIGURE 1.4
360^0 Video Viewing on VR Headset

The viewport is the current region seen by the user at any moment in time. Compared to the traditional video content delivery as described above, the 360^0 video applications require higher bandwidth and lower latency requirements that put pressure on the underlying networks. In order to be able to accommodate these types of applications and increase their adoption, various design solutions have been proposed in the literature especially when delivered over dynamic wireless networks environments. Some of the solutions involve streaming the entire 360^0 video content to the users. However, this results in wasting the network resources. Consequently, another approach seen in the literature is to only transmit the content within or surrounding the viewport. In this way, the bandwidth requirements are reduced but the computational resources to enable this process are increased. Moreover, as 360^0 video requires a very high video resolution (i.e., larger than 4K) its delivery over a dynamic heterogeneous wireless environments is highly challenging especially while maintaining a high user QoE.

1.2.3 360^0 Mulsemedia Streaming Applications

The next generation of wireless networks like 5G and beyond along with the advancements in mobile devices that are becoming more and more powerful, enables a new range of innovative applications that could improve the multimedia experience of the mobile users. One of these applications is represented by 360^0 MULtiple SEnsorial MEDIA (Mulsemedia) which enriches the 360° video content with other media objects like olfactory, haptic or even thermoceptic ones [46]. This type of application could revolutionize the streaming technology by creating a realistic experience for the mobile users [27]. An illustration of a possible Mulsemedia setup is represented in Figure 1.5.

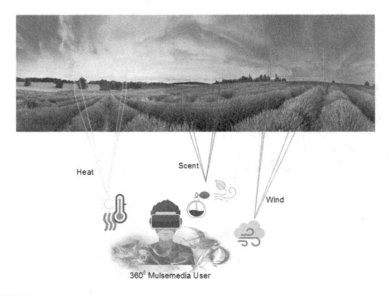

FIGURE 1.5
360^0 Mulsemedia Experience Scenario

The HMD is enhanced with diffuse capabilities for olfaction, air-flow (wind-induced haptic effects) and thermoception/heat while the 360^0 video content is streamed from a web server. This new application category combines the advantages of conventional mulsemedia applications [47] with the delivery of interactive 360^0 video. However, this realistic virtual world enhanced with additional sensorial effects (i.e. thermoception, olfaction and haptic) comes at the cost of stringent QoS requirements, representing a challenge even for the highly anticipated 5G and beyond networks [48].

An example of adaptive 360^0 mulsemedia delivery over 5G networks is illustrated in Figure 1.6 [48]. The scenario envisions a 360^0 mulsemedia user accessing content from a 360^0 mulsemedia server over the 5G network. The 360^0 mulsemedia content will be streamed in an adaptive manner, over the future 5G network to mobile terminals or HMDs enhanced with diffuser capabilities for olfaction, air-flow (wind-induced haptic effects) and thermoception/heat. In this way, the end-user can experience different intensities of sensorial components according to the viewing direction within the 360^0 panoramic video which could increase their QoE levels [49; 50]. However, due to stringent latency and bandwidth requirements of this application class [48], the provisioning of high performance QoE will be a challenge even for the highly anticipated 5G networks. Thus, it is foreseen that the mobile operators will have to deal with an explosion of mobile broadband data traffic over their networks. This puts significant pressure on them, especially as they will need to accommodate increasing users' demands as well as meeting their QoE expectations.

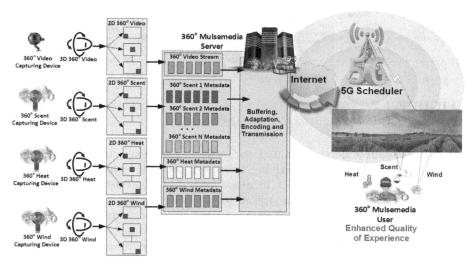

FIGURE 1.6
360^0 Mulsemedia Delivery System

Consequently, more advanced solutions need be adopted to be able to maintain acceptable QoE levels for the end-users when consuming these new immersive multimedia-rich applications [27]. These innovative and advanced applications and services are know to be resource-hungry applications, with very high data rates and extremely low latency requirements. Thus, in the context of 5G/6G, classical radio resource management functionalities might not be able to meet the stringent QoS requirements. In this context, the integration of ML-based solution is currently gaining considerable attention for resource allocation and QoS provisioning [51–53] as well as for the development of intelligent services in highly dynamic networks [48; 54; 55].

1.2.4   QoS and QoE Provisioning over Wireless Networks

As QoE is predicted to become the biggest differentiator between network operators [18], it is important to integrate advanced solutions that could cater for these new immersive multimedia-rich applications in terms of QoS requirements [27]. Consequently, when dealing with rich multimedia content delivery, two important concepts that need to be defined are QoS and QoE as illustrated in Figure 1.7.

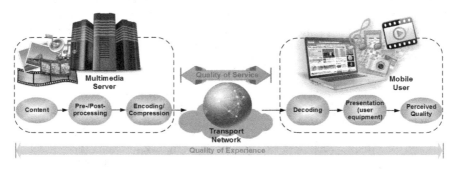

FIGURE 1.7
QoS vs. QoE in Multimedia Content Delivery

QoS measures network-related parameters, like delay, jitter, packet loss, etc. as provided by the underlying transport network technology. However, QoS does not a direct impact in guaranteeing the end-use satisfaction. In this regard, QoE relates to the overall service performance as being perceived subjectively by the end-user. This aspect is better represented in Figure 1.8 which illustrates the wide range of factors that contribute to the overall QoE of a mobile user within a heterogeneous wireless environment. It is noted that the overall QoE of the mobile user is influenced by the entire end-to-end system starting from the Operator (e.g., different pricing models for various class of services, etc.) up to the Culture (e.g., religion, economics, social class, etc.) [27].

Enabling QoS provisioning for rich media services, under a dynamic and complex heterogeneous wireless environment (e.g., moving people, objects, etc.) presents great challenges [48; 56]. Table 1.1 presents eight user-centric QoS classes for a range of multimedia-based services and applications as defined by the ITU-T Recommendation G.1010 [1] based on the delay range and loss sensitivities.

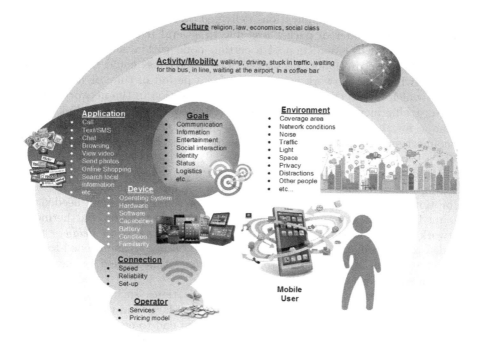

FIGURE 1.8

Parameters that Impact the Mobile User Experience

1.2.5 QoS Provisioning for Multimedia Delivery

The QoS parameters that have the most impact on the QoE of multimedia applications are packet loss ration, delay, jitter (delay variations) and

TABLE 1.1

The G.1010 Model for User-Centric QoS Categories [1]

	Error Tolerant	Error Intolerant
Interactive (delay < 1 sec)	Conventional Voice and Video	Command/control (e.g., Telnet, interactive games)
Responsive (delay ≈ 2 sec)	Voice and video messaging	Transactions (e.g., eCommerce, Web browsing, e-mail access)
Timely (delay ≈ 10 sec)	Streaming audio and video	Messaging and downloads (e.g., FTP, still images)
Non-critical (delay>10 sec)	Fax	Background (e.g., Usenet)

bandwidth. Multimedia applications delivered over a mobile and wireless environment could suffer of packet loss due to several reasons: (1) congestion – the limited-queue-exhaustion of a certain node within the network could lead to the node dropping the packet. The packet loss could be either distributed where the network is congested for a period of time or it could be bursty, which means that there is a sudden congestion in the network due to a short traffic increase; (2) network errors – due to nosy links or link-errors common for the wireless environment (drop in signal strength, wireless link disconnections, etc.) the packet might be marked as corrupted and discarded [26].

The delay consists of several parts: *end-point application delay* representing the time difference between the arrival of the media content and the consumption of media content; *network delay* representing the time needed for the media content to travel from the source to the destination. Furthermore, network delay consists of three parts, such as: *transmission delay* – time needed to transmit the packet; *packet processing delay* – time needed to process a packet, e.g., queuing; and *propagation delay* – time needed for the packet to reach the destination [27].

In the case of interactive multimedia applications, such as gaming, immersive applications or video conferencing, a very large end-to-end delay will have a significant impact on QoE. Whereas, the end-to-end delay for Video on Demand (VoD) applications might not have a significant impact on QoE because the data packets can get relatively offset on the network transport path regardless of the delay conditions [26; 57].

The delay variation caused by network congestion, queuing delays, processing delays, signal drop, path changes or other reasons is referred to as *jitter*. Jitter can be avoided by implementing a buffer at the receiver that will store a certain amount of packets before sending them to the decoder. However, in case the buffer is full or if the packets arrive slowly, it might cause degradation in the users' QoE. However, nowadays the direct impact of the jitter on the video-based application can be neglected due to the increase in buffer sizes.

Another important parameter is the bandwidth availability that could also be considered a contributing factor when determining the performance of the end user's application. As we have seen previously, various applications have different QoS requirements. For example, one could estimate the completion time of a file transfer by knowing the available bandwidth. Thus, we could conclude that quality of multimedia-based applications is dependent on a wide range of parameters: the mobile device capabilities and characteristics, the type of Radio Access Technology (RAT), the application requirements, and network conditions.

1.2.6   Approaches for Measuring the Video Quality

Different methodologies were developed when trying to assess the end-user perceived video quality levels when streaming over dynamic wireless networks. Two main categories are defined: *Subjective methods* and *Objective methods* [27].

The subject methods are more reliable because they are performed directly on human subjects. However, they are time consuming and have a high implementation cost. Thus, this makes them unsuitable when dealing with real time assessment.

On the other side, the objective methods are further classified into three subgroups [27]:

- *full reference methods* – these methods compare the original video with the distorted one, making them more precise and more correlated with the subjective methods. However, the computational complexity is increased significantly as they rely on per-pixel processing and synchronization between the original video and the distorted one. Some examples of typically used metrics in the full reference methods are: blockiness, blur, brightness, contrast, jerkiness, frame skips, freezes, etc. [58].

- *reduced reference methods* – these methods are extracting specific features from the original video. These features are transmitted to the receiver where the same information is extracted from the received distorted video and compared with the ones of the original video.

- *no reference methods* – these methods are not dependent on the original video. Complex algorithms are used and applied on the distorted signal only. This makes them more applicable with less computational complexity and could actually be used in analyzing live streams [26]. Some examples of proprietary no reference methods include: Video Streaming Quality Index (VSQI) [59], Mobile TV Quality Index (MTQI), Video Telephony Quality Index (VTQI) and Perceptual Evaluation of Video Quality (PEVQ). Typical used metrics considered by the proprietary solutions include: video codec used, total bit rate, duration of initial buffering, number and duration of re-buffering periods and packet loss, etc.

However, one of the most important metrics used when assessing the end-user perceived video quality level is the Mean Opinion Score (MOS) [58]. The typical five MOS levels used for describing the quality and impairment of a multimedia stream are illustrated in Table 1.2 [58], where Level 1 represents *Bad* quality and Level 5 represents *Excellent* quality.

TABLE 1.2
The Mean Opinion Score Levels

MOS	Quality	Impairment
5	Excellent	Imperceptible
4	Good	Perceptible but not annoying
3	Fair	Slightly annoying
2	Poor	Annoying
1	Bad	Very Annoying

One of the most common metrics for objective video quality assessment is Peak Signal to Noise Ratio (PSNR) given by eq. 1.1 [26].

$$PSNR_{dB} = 20\log_{10}\frac{255}{\sqrt{MSE}} \tag{1.1}$$

where, Mean Square Error (MSE) is defined as the cumulative squared error between the original and the processed video. There are different approaches for computing PSNR including definitions for online and offline use. Lee et al. [60] define PSNR as in eq. 1.2.

$$PSNR_{dB} = 20\log_{10}\frac{MAX\_Bitrate}{\sqrt{(EXP\_Thr - CRT\_Thr)^2}} \tag{1.2}$$

where $MAX\_Bitrate$ is the bitrate of the multimedia stream after the encoding process, EXP_Thr is the expected average throughput for the delivery of the multimedia stream over the network, and CRT_Thr is the actual average received throughput for the multimedia delivery over the network.

Significant efforts have been put into developing objective Video Quality Assessment (VQA) metrics to enable the automatic estimation of the video quality. A comprehensive survey on the VQA methods is presented in [61]. However, MOS still remains the most reliable measure of video quality when gathered directly from the users through subjective tests. To overcome these challenges, Moldovan et al. [62] propose a novel tool referred to as VQAMap, that takes as input the subjective data from public VQA databases and automatically creates mapping rules of any objective VQA metric to the MOS scale [18].

FIGURE 1.9
QoE for 360^0 Video Streaming

Compared to the conventional videos, when assessing the perceived QoE for 360^0 videos, there are several other factors that need to be taken into account as illustrated in Figure 1.9. It can be noted that apart from the visual/perceptual quality we also need to look into the depth quality as defined by the realism, power, presence, etc. as well as the fatigue, (dis)comfort or cybersickness involved in this type of applications. In this context, it is important that we first determine the user's specific quality aspects when dealing with 360^0 video content distribution. Zhang et al. [63] proposed SAMPVIQ, a subjective assessment method for panoramic videos that aims to overcome the performance gap between the quality evaluation metrics applied to 360^0 videos as compared to conventional videos. Additionally, Xu et al. [64] proposed two subjective quality evaluation metrics to assess the quality loss in 360^0 videos, such as: the overall differential mean opinion score (O-DMOS) and vectorized DMOS (V-DMOS). Moreover, some open source tools for 360^0 videos that are freely available are: AVTrack360 [65], OpenTrack and 360player [66] capturing the users' head traces while watching 360^0 videos, and VRate [67], a Unity-based tool that integrates subjective questionnaires in a VR environment. Furthermore, Pérez et al. [68] proposed MIRO360, a full-fledged Android-based application that facilitates the guidelines development for future VR subjective tests. However, one significant barrier in achieving high QoE levels for some users, especially when delivering 360^0 videos, is represented by cybersickness, that can cause fatigue, nausea, discomfort and aversion [69].

1.3 A Simplified Network Model

A simplified communication model represented by a general block diagram is illustrated in Figure 1.10 [70; 71].

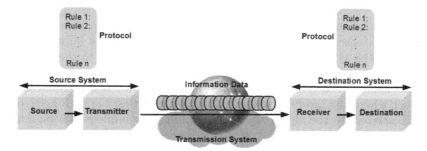

FIGURE 1.10
Simplified Communication Model

The model consists of five main components, such as:

- Information Data – representing the message to be communicated, e.g., text, audio, video, etc.

- Source System – consists of a *Source* and a *Transmitter*, where the source generates the data to be transmitted and the transmitter transforms and encodes the data to be transmitted in the form of electromagnetic signals across a transmission system. The source system can be a computer, server, handheld device, etc.

- Transmission System – this represents the physical path the data takes to travel from the source to the destination. It can be represented by wired, wireless links or even a more complex heterogeneous network that connects the source with the destination.

- Destination System – consists of *Receiver* and *Destination*, where the receiver will accept the signal received over the transmission system and will convert it in the right format for the destination device. Similarly, the destination system can be a computer, server, handheld device, etc.

- Protocol – represents a set of rules that enables the communication between devices. For two devices to be able to communicate they need to *speak the same language*. Thus, the protocol represents the mutually accepted conventions between the two communicating devices that enables their communication. To this end, the key elements of a protocol are: (1) syntax – refers to the structure or format of the data, as well as signal levels, (2) semantics – refers to the interpretation of each section of bits or fields within the data structure and (3) timing – refers to the speed matching and sequencing of the communicating devices.

To enable the communication between devices across complex dynamic networks/systems, a layered approach was adopted as developed by International Standards Organisation (ISO). The layered approach brings several advantages, including: enables a good identification and definition of the system components, enables the standardization of the relationship between system components, eases maintenance and system update, enables implementation changes at the system components level to be transparent to the rest of the system, etc.

An example of a general layer-based network model is illustrated in Figure 1.11 where the layers have the following properties: (1) each layer offers certain *services* to the higher layers while hiding the implementation details, enabling *modularity*; (2) any layer in one network node will communicate with the corresponding layer in another node by the means of a protocol; (3) *interfaces* between adjacent layers define which operations and services the lower layer offers to the higher layer, minimizing the amount of information passed between layers.

Consequently, the basic components of a layered architecture are protocol stack, protocol, service and interface. The *protocol stack* is actually represented by the layered architecture of protocols that are used for managing the data exchanged between devices. The *service* represents the set of operations a layer will provide to the higher layer and *interface* represents the way through which the data is transferred from one layer to another layer.

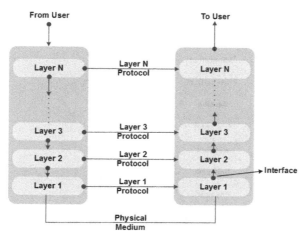

FIGURE 1.11
Layer-Based Network Model

Two of the most well-known reference models for a layered architecture are the Open Systems Interconnection (OSI) [72] reference model and the Transmission Control Protocol (TCP)/Internet Protocol (IP) [73] protocol suite as illustrated in Figure 1.12.

The OSI model was developed by ISO in late 1970s to facilitate the communication between different types of computer systems. The OSI model consists of seven layers with each layer performing different function, such as:

- **layer 1 – physical** – deals with the physical characteristics of the interfaces and the transmission medium. It is also responsible with the transmission of unstructured **bit streams** from one node to the next.

- **layer 2 – data link** – enables the reliable transmission over the physical link and divides the stream of bits into data units called **frames** with support for synchronization, error control and flow control.

- **layer 3 – network** – deals with the delivery of **packets** between any source-to-destination pairs by means of logical addressing and routing. It is also responsible for establishing, maintaining and terminating connections.

- **layer 4 – transport** – offers transparent transfer of data **segments** between end points, and provides end-to-end flow control and error correction.

- **layer 5 – session** – establishes, maintains, synchronizes and terminates connections between communicating applications.

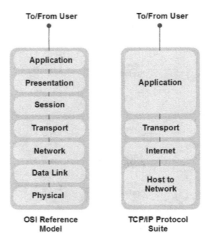

FIGURE 1.12
OSI Refernce Model vs. TCP/IP Reference Model

- **layer 6 – presentation** – deals with the translation, compression and encryption of data.

- **layer 7 – application** – enables the users to access the provided services.

The TCP/IP protocol suite was initially defined as having four layers, such as: host to network, internet, transport and application. As illustrated in Figure 1.12, the host to network layer corresponds to the combined physical and data link layers of the OSI, the internet layer corresponds to the network layer of the OSI, while the application layer is equivalent to the combination of session, presentation and application layers of the OSI. Each layer in the TCP/IP protocol suite consists of independent protocols that could be mixed and matched based on the system requirements. Whereas in the OSI model, each layers has well-defined functions. Both OSI reference model and the TCP/IP protocol suite have been criticized in terms of the concept with the OSI model being too theoretical while the TCP/IP being too practical, in terms of protocols with the TCP/IP consisting of very popular protocols while the OSI protocol are less known, and in terms of layers with TCP/IP having too few while OSI having too many. Consequently, the compromise was a five-layer hybrid model as illustrated in Figure 1.13 consisting of: physical, data link, network, transport and application layers [3].

In this simple wireless communication system as illustrated in Figure 1.13 the layers are responsible for: (1) *the physical layer* deals with frequency selection, generation of carrier frequency, signal detection, modulation of data onto a carrier frequency and encryption; (2) *the data link layer* deals with medium access, multiplexing, detection and correction of transmission errors and

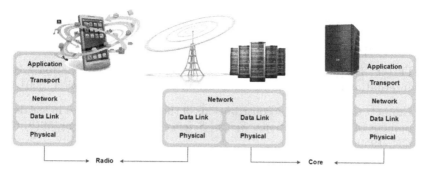

FIGURE 1.13
Five-Layer Reference Model for a Simple Wireless Communication System –
Example

detection of data frames; (3) *the network layer* – deals with routing of packets through the network, addressing, handover between different networks. E.g., IP; (4) *the transport layer* deals with QoS, flow and congestion control. E.g., TCP, and (5) *the application layer* consists of the protocols designed for fulfilling communication needs of an application. E.g., Hypertext Transfer Protocol (HTTP).

The system as exemplified in Figure 1.13 consists of a mobile user communicating with a fixed user over a mobile network. It can be noted that the protocol stack is implemented in the system as per the reference model. The end systems, in this case the mobile device and the computer of the end-users would require the full implementation of the protocol stack including all five layers. However, the intermediate systems, connecting the two end system do not necessarily need all five layer, but only the three layers to enable the communication.

The use of standards has been widely accepted by the communications industry as a way to enable the national and international interoperability of systems made by different manufacturers. The main organizations that deal with the creation of legal as well as informal standards are listed below:

- International Standards Organization (ISO)

- International Telecommunication Union-Telecommunication Standards Sector (ITU-T)

- American National Standards Institute (ANSI)

- International Electrotechnical Commission (IEC)

- Federal Communications Commission (FCC)

- Internet Engineering Task Force (IETF)

- Institute of Electrical and Electronics Engineers (IEEE)

Bibliography

[1] ITU-T. End-user Multimedia QoS Categories. Recommendation g.1010, International Telecommunication Union – Telecommunication, 2001.

[2] Emmett C Belzer. Some contributions of the bell laboratories in the development of communications. *School Science and Mathematics*, 50(8):652–654, 1950.

[3] Jochen H Schiller. *Mobile communications*. Pearson education, 2003.

[4] Mark Weiser. The computer for the 21st century. *ACM SIGMOBILE Mobile Computing and Communications Review*, 3(3):3–11, 1999.

[5] Friedemann Mattern and Christian Floerkemeier. From the internet of computers to the internet of things. In *From active data management to event-based systems and more*, pages 242–259. Springer, 2010.

[6] Ismael Peña-López et al. Itu internet report 2005: the internet of things. 2005.

[7] Irena Bojanova, George Hurlburt, and Jeffrey Voas. Imagineering an internet of anything. *Computer*, 47(6):72–77, 2014.

[8] Ioan-Sorin Comşa, Andreea Molnar, Irina Tal, Per Bergamin, Gabriel-Miro Muntean, Cristina Hava Muntean, and Ramona Trestian. A machine learning resource allocation solution to improve video quality in remote education. *IEEE Transactions on Broadcasting*, 67(3):664–684, 2021.

[9] Robert Connor Chick, Guy Travis Clifton, Kaitlin M Peace, Brandon W Propper, Diane F Hale, Adnan A Alseidi, and Timothy J Vreeland. Using technology to maintain the education of residents during the covid-19 pandemic. *Journal of Surgical Education*, 77(4):729–732, 2020.

[10] Pol Pérez Sust, Oscar Solans, Joan Carles Fajardo, Manuel Medina Peralta, Pepi Rodenas, Jordi Gabaldà, Luis Garcia Eroles, Adrià Comella, César Velasco Muñoz, Josuè Sallent Ribes, et al. Turning the crisis into an opportunity: digital health strategies deployed during the covid-19 outbreak. *JMIR Public Health and Surveillance*, 6(2):e19106, 2020.

[11] Yushan Siriwardhana, Gürkan Gür, Mika Ylianttila, and Madhusanka Liyanage. The role of 5g for digital healthcare against covid-19 pandemic: Opportunities and challenges. *ICT Express*, 7(2):244–252, 2021.

[12] Mohsin Raza, Priyan Malarvizhi Kumar, Dang Viet Hung, William Davis, Huan Nguyen, and Ramona Trestian. A digital twin framework for industry 4.0 enabling next-gen manufacturing. In *2020 9th International*

Conference on Industrial Technology and Management (ICITM), pages 73–77. IEEE, 2020.

[13] Huan X Nguyen, Ramona Trestian, Duc To, and Mallik Tatipamula. Digital twin for 5g and beyond. *IEEE Communications Magazine*, 59(2):10–15, 2021.

[14] Zhengquan Zhang, Yue Xiao, Zheng Ma, Ming Xiao, Zhiguo Ding, Xianfu Lei, George K Karagiannidis, and Pingzhi Fan. 6g wireless networks: Vision, requirements, architecture, and key technologies. *IEEE Vehicular Technology Magazine*, 14(3):28–41, 2019.

[15] The 5G Infrastructure Association. European vision for the 6g network ecosystem. 2021.

[16] Ping Yang, Yue Xiao, Ming Xiao, and Shaoqian Li. 6g wireless communications: Vision and potential techniques. *IEEE Network*, 33(4):70–75, 2019.

[17] Ramona Trestian and Gabriel-Miro Muntean. *Paving the way for 5G through the convergence of wireless systems*. IGI Global, 2018.

[18] Ramona Trestian, Ioan-Sorin Comsa, and Mehmet Fatih Tuysuz. Seamless multimedia delivery within a heterogeneous wireless networks environment: Are we there yet? *IEEE Communications Surveys & Tutorials*, 20(2):945–977, 2018.

[19] VNI Cisco et al. Cisco visual networking index: Forecast and trends 2018-2023 white paper. 2019.

[20] Ramona Trestian, Quoc-Tuan Vien, Huan X Nguyen, and Orhan Gemikonakli. Eco-m: Energy-efficient cluster-oriented multimedia delivery in a lte d2d environment. In *2015 IEEE International Conference on Communications (ICC)*, pages 55–61. IEEE, 2015.

[21] L. Zou, R. Trestian, and G. Muntean. E3doas: Balancing qoe and energy-saving for multi-device adaptation in future mobile wireless video delivery. *IEEE Transactions on Broadcasting*, 64(1):26–40, 2018.

[22] Longhao Zou, Ramona Trestian, and Gabriel-Miro Muntean. E 2 doas: user experience meets energy saving for multi-device adaptive video delivery. In *2015 IEEE Conference on Computer Communications Workshops (INFOCOM WKSHPS)*, pages 444–449. IEEE, 2015.

[23] Pasquale Scopelliti, Angelo Tropeano, Gabriel-Miro Muntean, and Giuseppe Araniti. An energy-quality utility-based adaptive scheduling solution for mobile users in dense networks. *TBC*, 66(1):47–55, 2020.

[24] Ioan-Sorin Comșa, Ramona Trestian, Gabriel-Miro Muntean, and Gheorghiță Ghinea. 5mart: A 5g smart scheduling framework for optimizing qos through reinforcement learning. *IEEE Transactions on Network and Service Management*, 17(2):1110–1124, 2019.

[25] Adriana Hava, Yacine Ghamri-Doudane, John Murphy, and Gabriel-Miro Muntean. A load balancing solution for improving video quality in loaded wireless network conditions. *TBC*, 65(4):742–754, 2019.

[26] Ramona Trestian. *User-centric power-friendly quality-based network selection strategy for heterogeneous wireless environments*. PhD thesis, Dublin City University, 2012.

[27] Ramona Trestian and Gabriel-Miro Muntean. Solutions for improving rich media streaming quality in heterogeneous network environments. 2021.

[28] Microsoft. IIS Smooth Streaming Technical Overview. 2010.

[29] R. Trestian, Q. Vien, P. Shah, and G. Mapp. Uefa-m: Utility-based energy efficient adaptive multimedia mechanism over lte hetnet small cells. In *2017 International Symposium on Wireless Communication Systems (ISWCS)*, pages 408–413, 2017.

[30] R. Trestian, G. Muntean, and O. Ormond. Signal strength-based adaptive multimedia delivery mechanism. In *2009 IEEE 34th Conference on Local Computer Networks*, pages 297–300, 2009.

[31] R. Trestian, Q. Vien, H. X. Nguyen, and O. Gemikonakli. Eco-m: Energy-efficient cluster-oriented multimedia delivery in a lte d2d environment. In *2015 IEEE International Conference on Communications (ICC)*, pages 55–61, 2015.

[32] M. Kennedy, H. Venkataraman, and G. Muntean. Battery and stream-aware adaptive multimedia delivery for wireless devices. In *IEEE Local Computer Network Conference*, pages 843–846, 2010.

[33] Wanxin Shi, Qing Li, Ruishan Zhang, Gengbiao Shen, Yong Jiang, Zhenhui Yuan, and Gabriel-Miro Muntean. Qoe ready to respond: A qoe-aware mec selection scheme for dash-based adaptive video streaming to mobile users. In *Proceedings of the 29th ACM International Conference on Multimedia*, pages 4016–4024, 2021.

[34] A. Molnar and C. H. Muntean. Cost-oriented adaptive multimedia delivery. *IEEE Transactions on Broadcasting*, 59(3):484–499, 2013.

[35] A. Molnar and C. H. Muntean. Comedy: Viewer trade off between multimedia quality and monetary benefits. In *2013 IEEE International Symposium on Broadband Multimedia Systems and Broadcasting (BMSB)*, pages 1–7, 2013.

[36] Andreea Molnar and Cristina Hava Muntean. Assessing learning achievements when reducing mobile video quality. *J. UCS*, 21(7):959–975, 2015.

[37] G. Muntean, P. Perry, and L. Murphy. Objective and subjective evaluation of qoas video streaming over broadband networks. *IEEE Transactions on Network and Service Management*, 2(1):19–28, 2005.

[38] A. N. Moldovan and C. H. Muntean. Dqamlearn: Device and qoe-aware adaptive multimedia mobile learning framework. *IEEE Transactions on Broadcasting*, 99:1–16, 2020.

[39] M. Anedda, M. Murroni, and G. Muntean. A novel markov decision process-based solution for improved quality prioritized video delivery. *IEEE Transactions on Network and Service Management*, 17(1):592–606, 2020.

[40] M. Kennedy, H. Venkataraman, and G. Muntean. Dynamic stream control for energy efficient video streaming. In *2011 IEEE International Symposium on Broadband Multimedia Systems and Broadcasting (BMSB)*, pages 1–6, 2011.

[41] Hrishikesh Venkataraman, Poornachand Kalyampudi, and Gabriel-Miro Muntean. Cashew: Cluster-based adaptive scheme for multimedia delivery in heterogeneous wireless networks. *Wireless Personal Communications*, 62(3):517–536, 2012.

[42] G.-M. Muntean and N. Cranley. Resource Efficient Quality-Oriented Wireless Broadcasting of Adaptive Multimedia Content. *IEEE Transactions on Broadcasting*, 53(1):362–368, March 2007.

[43] G. M. Muntean. Efficient delivery of multimedia streams over broadband networks using qoas. *IEEE Transactions on Broadcasting*, 52(2):230–235, 2006.

[44] Ramona Trestian, Gabriel-Miro Muntean, and Olga Ormond. Signal strength-based adaptive multimedia delivery mechanism. In *2009 IEEE 34th Conference on Local Computer Networks*, pages 297–300. IEEE, 2009.

[45] A. Yaqoob, T. Bi, and G. M. Muntean. A survey on adaptive 360° video streaming: Solutions, challenges and opportunities. *IEEE Communications Surveys Tutorials*, 22(4):2801–2838, 2020.

[46] I.-S. Comsa, E. B. Saleme, A. Covaci, G. M. Assres, R. Trestian, C. A. S. Santos, and G. Ghinea. Do i smell coffee? the tale of a 360° mulsemedia experience. *IEEE MultiMedia*, 27(1):27–36, 2020.

[47] Niall Murray. Mulsemedia: multiple sensorial media – towards understanding user and quality of experience of future media applications. In

Joel A. F. dos Santos and Débora Christina Muchaluat-Saade, editors, *WebMedia*, page 3. ACM, 2019.

[48] I.-S. Comsa, R. Trestian, and G. Ghinea. 360° mulsemedia experience over next generation wireless networks – a reinforcement learning approach. In *2018 Tenth International Conference on Quality of Multimedia Experience (QoMEX)*, pages 1–6, 2018.

[49] Estêvão B Saleme, Alexandra Covaci, Gebremariam Assres, Ioan-Sorin Comsa, Ramona Trestian, Celso AS Santos, and Gheorghita Ghinea. The influence of human factors on 360° mulsemedia qoe. *International Journal of Human-Computer Studies*, 146:102550, 2021.

[50] Alexandra Covaci, Ramona Trestian, Estêvão Bissoli Saleme, Ioan-Sorin Comsa, Gebremariam Assres, Celso AS Santos, and Gheorghita Ghinea. 360 mulsemedia: a way to improve subjective qoe in 360 videos. In *Proceedings of the 27th ACM International Conference on Multimedia*, pages 2378–2386, 2019.

[51] Ioan-Sorin Comsa, Sijing Zhang, Mehmet Aydin, Pierre Kuonen, Ramona Trestian, and Gheorghita Ghinea. Enhancing user fairness in ofdma radio access networks through machine learning. In *Wireless Days*, pages 1–8. IEEE, 2019.

[52] I.-S. Comsa, S. Zhang, M. E. Aydin, P. Kuonen, Y. Lu, R. Trestian, and G. Ghinea. Towards 5g: A reinforcement learning-based scheduling solution for data traffic management. *IEEE Transactions on Network and Service Management*, 15(4):1661–1675, 2018.

[53] A. Al-Jawad, P. Shah, O. Gemikonakli, and R. Trestian. Learnqos: A learning approach for optimizing qos over multimedia-based sdns. In *2018 IEEE International Symposium on Broadband Multimedia Systems and Broadcasting (BMSB)*, pages 1–6, 2018.

[54] I.-S. Comsa, R. Trestian, G. Muntean, and G. Ghinea. 5mart: A 5g smart scheduling framework for optimizing qos through reinforcement learning. *IEEE Transactions on Network and Service Management*, 17(2):1110–1124, 2020.

[55] I.-S. Comsa, G. Muntean, and R. Trestian. An innovative machine-learning-based scheduling solution for improving live uhd video streaming quality in highly dynamic network environments. *IEEE Transactions on Broadcasting*, pages 1–13, 2020.

[56] Subha Dhesikan and Seong-Ho Jeong. Itu-t sg16 qos architecture for multimedia systems. In *ICAS/ICNS*, page 61. IEEE Computer Society, 2005.

[57] Mathieu Carnec, Patrick Le Callet, and Dominique Barba. Full reference and reduced reference metrics for image quality assessment. In *ISSPA (1)*, pages 477–480. IEEE, 2003.

[58] ITU-T. Subjective Video Quality Assessment Methods for Multimedia Applications. Recommendation p.910, International Telecommunication Union - Telecommunication, 1999.

[59] Ascom. Video Streaming Quality Measurement with VSQI. TEMS, 2009.

[60] S. B. Lee, G. M. Muntean, and A. F. Smeaton. Performance-aware replication of distributed pre-recorded iptv content. *IEEE Transactions on Broadcasting*, 55(2):516–526, June 2009.

[61] S. Chikkerur, V. Sundaram, M. Reisslein, and L. J. Karam. Objective video quality assessment methods: A classification, review, and performance comparison. *IEEE Transactions on Broadcasting*, 57(2):165–182, June 2011.

[62] A. N. Moldovan, I. Ghergulescu, and C. H. Muntean. Vqamap: A novel mechanism for mapping objective video quality metrics to subjective mos scale. *IEEE Transactions on Broadcasting*, 62(3):610–627, Sept 2016.

[63] Bo Zhang, Junzhe Zhao, Shu Yang, Yang Zhang, Jing Wang, and Zesong Fei. Subjective and objective quality assessment of panoramic videos in virtual reality environments. In *ICME Workshops*, pages 163–168. IEEE Computer Society, 2017.

[64] Mai Xu, Chen Li, Zhenzhong Chen, Zulin Wang, and Zhenyu Guan. Assessing visual quality of omnidirectional videos. *IEEE Trans. Circuits Syst. Video Techn.*, 29(12):3516–3530, 2019.

[65] Stephan Fremerey, Ashutosh Singla, Kay Meseberg, and Alexander Raake. Avtrack360: an open dataset and software recording people's head rotations watching 360° videos on an hmd. In Pablo César, Michael Zink, and Niall Murray, editors, *MMSys*, pages 403–408. ACM, 2018.

[66] Xavier Corbillon, Francesca De Simone, and Gwendal Simon. 360-degree video head movement dataset. In *Proceedings of the 8th ACM on Multimedia Systems Conference*, MMSys'17, page 199–204, New York, NY, USA, 2017. Association for Computing Machinery.

[67] Georg Regal, Raimund Schatz, Johann Schrammel, and Stefan Suette. Vrate: A unity3d asset for integrating subjective assessment questionnaires in virtual environments. In *QoMEX*, pages 1–3. IEEE, 2018.

[68] Pablo Pérez and Javier Escobar. Miro360: A tool for subjective assessment of 360 degree video for itu-t p.360-vr. In *QoMEX*, pages 1–3. IEEE, 2019.

[69] Ajoy S. Fernandes and Steven K. Feiner. Combating vr sickness through subtle dynamic field-of-view modification. In Bruce H. Thomas, Robert Lindeman, and Maud Marchal, editors, *3DUI*, pages 201–210. IEEE Computer Society, 2016.

[70] William Stallings. *Data and Computer Communications*. Prentice-Hall, Englewood Cliffs, New Jersey, 6th edition, September 1999.

[71] Behrouz A. Forouzan. *Data Communications and Networking*. McGraw-Hill, Inc., USA, 3 edition, 2003.

[72] John D Day and Hubert Zimmermann. The osi reference model. *Proceedings of the IEEE*, 71(12):1334–1340, 1983.

[73] Christoph Meinel and Harald Sack. The foundation of the internet: Tcp/ip reference model. In *Internetworking*, pages 29–61. Springer, 2013.

Chapter 2

Wireless Transmission Fundamentals

"The world isn't getting any easier. With all these new inventions I believe that people are hurried more and pushed more. The hurried way is not the right way; you need time for everything - time to work, time to play, time to rest."

Hedy Lamarr
Wireless Visionary,
Inventor of the frequency hopping spread spectrum technology

2.1 Spectrum and Frequencies

Wireless communication deals with the transmission of electromagnetic waves over an unguided media, like the atmosphere or the outer space. A wireless transmission can use different frequency bands with each having certain advantages and disadvantages. The electromagnetic spectrum as illustrated in Figure 2.1 gives an overview of a range of sources of radiating devices and their corresponding frequencies and wavelengths. The relationship between the frequency and the wavelength is given by equation 2.1 [1].

$$\lambda = c/f \qquad (2.1)$$

where $c = 3 \times 10^8$m/s represents the speed of light in vacuum, f is the frequency in Hertz and λ is the wavelength in meters.

From Figure 2.1, it can be noted that as the frequency increases, the wavelength gets shorter and the cycles get faster creating more energy that might disrupt the atom. The electromagnetic energy travels in waves consisting of photons with extremely low energies up to gamma-rays the highest marking the extremes of the electromagnetic spectrum. Within the two extremes we have the Radio Frequency (RF), microwaves, infrared light, visible light, ultraviolet light and x-rays. However, when the radiation is powerful enough to disrupt the atom and break the chemical bonds, it is called ionizing radiation. When the radiation is not powerful enough, it is called non-ionizing radiation. The division between the two types of radiation is happening within the ultraviolet range.

DOI: 10.1201/9781003222095-2

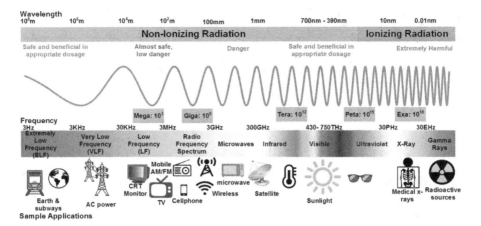

FIGURE 2.1
The Electromagnetic Spectrum

Consequently, it can be noted that the gamma rays, x-rays and some ultraviolet waves belong to the ionizing radiation type, meaning that their impact could be extremely harmful for humans, for example sunburns. However, sometimes they are used in a positive way, for example exposure to radiation to kill cancer cells.

Within some regions, the electromagnetic radiation is impacted by issues like atmospheric and water absorption, ambient sunlight creating these windows of opaque atmosphere to wavelengths and limiting their use for wireless communication. On the other side, the atmospheric windows that consist of the spectrum regions where the wavelengths can pass through the atmosphere are suitable for radio communication.

While some of the current wireless and mobile technologies like 5G are designed to use the mmWave spectrum (30 GHz to 300 GHz) to be able to enable data rates of up to 100 Gbps, the anticipated 6G networks are expected to operate within the 100 GHz to 1 THz frequency bands for even greater data rates [2].

In order to avoid interference, all the radio frequencies are regulated and this is done by different global regulatory bodies and standard agencies, like: Federal Communications Commission (FCC), European Telecommunication Standards Institute (ETSI), International Telecommunication Union (ITU), etc. The frequency bands are assigned based on three types of allocation [3]:

- licensed – the user needs a license to be able to transmit. Same frequency could be allocated to several users if they operate in non-overlapping areas.

- unlicensed – any user can transmit over the open spectrum bands, such as: unlicensed Industrial, Scientific and Medical (ISM) bands, unlicensed

FIGURE 2.2
Basic Wireless Communication System

ultra-wideband band, amateur radio frequency allocation. However, the users need to maintain the communication within certain thresholds, for example certain transmission power limits.

- not for use – frequency bands where no user can transmit as these are reserved for special use. For example, radio astronomy.

2.2 Signals for Conveying Information

Electromagnetic waves are used for conveying information from a source to a destination over a wireless transmission medium, via propagation through air, vacuum or sea water. An example of a basic wireless communication system is illustrated in Figure 2.2 and consists of a transmitter, the antennas radiating the electromagnetic energy into the transmission medium (e.g., the air) and the receiver. In most of the cases the same device can be a transmitter and a receiver, referred to as *transceiver*.

2.2.1 Basic Concepts and Terminology

This subsection will introduce some basic concepts and terminologies for data, signals and transmissions [1; 3–5].

Data represents the entity that conveys meaning or information. Data can be analog or digital. In the case of analog data, the information is continuous over some interval, e.g., voice and video. In the case of digital data, the information takes on discrete values, e.g., text and integers. Both analog and digital data can be propagate by analog or digital signals.

Signals are electric or electromagnetic encoding of data. Similarly to data, also signals can be analog or digital. Analog signals are represented by continuously varying electromagnetic waves, while digital signals are represented by

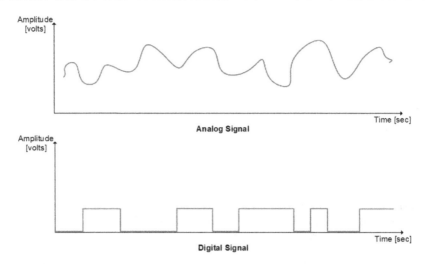

FIGURE 2.3
Analog and Digital Signals

sequences of voltage pulses. Analog data can be represented by analog or digital signals. For example, an analog wave is created in the air when someone speaks. This can be captured by a microphone and converted into an analog signal or sampled and converted into a digital signal. In the case of digital data, this could also be converted into a digital signal or modulated into an analog signal. An example of analog and digital signals is illustrated in Figure 2.3 [1].

Transmission represents the propagation of data and processing of signals. Similarly to data and signals, the transmission can also be analog or digital. The analog transmission can be used to transmit analog signals that could carry analog data or digital data. While digital transmission can be used to transmit both analog and digital signal, taking into account the content of the signal [4].

However, due to the imperfection of the transmission medium, the signal can suffer from attenuation, distortion or noise [5]. Thus, the receiver may receive a different signal from what it was transmitted. Signal **attenuation** represents the loss in energy of a signal as it travels through the medium. Signal **distortion** represents the change in the signal's shape or form as it travels through the medium. **Noise** represents the unwanted random disturbance that combines with the original transmitted signal as it travels through the medium.

In the case of analog transmission, the signal attenuation increases with the increase in the distance between the transmitter and the receiver. To overcome this aspect and achieve longer distances, amplifiers could be used.

However, the amplifier will just boost the energy of the input signal regardless of its content. Thus, the amplifier will also boost the noise components. Consequently, the more amplifiers we add to increase the distance, the signal will become more and more distorted [1].

In the case of digital transmission, the repeaters are used to increase the communication distance between the source and destination. In contrast to the amplifiers, the repeater takes into account the content of the input signal, by recovering the pattern of zeros and ones and generating a new clean signal [4].

2.2.2 Time and Frequency Domain

The most fundamental analog signal can be represented by a simple periodic sine wave. A **periodic** signal is defined as a signal in which a pattern is completed within a measurable time frame, referred to as **a period**. That pattern is repeated over subsequent periods until the completion of one full pattern is reached, referred to as **a cycle** [5]. If the signal has no repeating pattern or cycle is referred to as **aperiodic** signal.

Three main parameters are involved when representing a sine wave, such as: *amplitude* (A), *frequency* (f) and *phase* (ϕ) [4].

The amplitude can be defined as the maximum value or strength of a signal over time and is usually measured in *Volts (V)*. Figure 2.4(a) illustrates

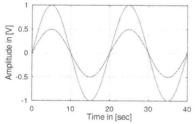

(a) Signals with different amplitudes

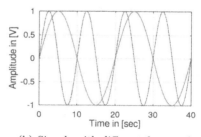

(b) Signals with different frequencies

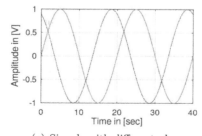

(c) Signals with different phases

FIGURE 2.4
Sine Wave Signal Parameters

an example of two signal with the same frequency and phase but different amplitude.

The frequency defines the periodicity of the signal and represents the rate in cycles per seconds at which the signal repeats. The frequency is usually measured in *Hertz (Hz)*. The relationship between the frequency f and the period T of the signal is given by equation 2.2.

$$T = 1/f \qquad (2.2)$$

Figure 2.4(a) illustrates an example of two signal with the same amplitude and phase but different frequency.

The phase can be defined as the measure of the signal's relative position in time within a single period. The phase is usually measured in *degrees*. Figure 2.4(a) illustrates an example of two signal with the same frequency and amplitude but different phase.

Consequently, we can use the three parameters to write down the general sine wave as a function of time, given in equation 2.3 and known as sinusoid [4].

$$s(t) = A\sin(2\pi ft + \phi) \qquad (2.3)$$

We have seen until now that sine waves are periodic signal whose parameters of amplitude, frequency and phase can be easily determined. Additionally, we have looked at the signal from the time-domain point of view which shows the signal's changes in amplitude over time. However, in practice, the electromagnetic signals are usually aperiodic and more complex, consisting of many different frequency components with each of it varying in amplitude. A graphical representation of a complex signal in the time domain is illustrated in Figure 2.5(a). This complex signal is in fact composed of five individual sine waves as illustrated in Figure 2.5(c). Each individual sine wave has a different frequency and different amplitude.

Fourier analysis can be used to show that every complex signal consists in fact of individual components with different amplitude, frequency and phase, and each of these components is in fact a sinusoid, and they are referred to as the fundamental frequency and a number of harmonic frequencies.

From Figure 2.5(a), it can be noted that it is quite difficult to determine the parameters of a complex signal in the time domain. For example, what would be the frequency components, and their respective power content of the signal as represented in Figure 2.5(a)? To better understand the relationship between the amplitude and frequency, we can look at this complex signals in the frequency domain as illustrated in Figure 2.5(b).

The frequency domain illustrates on the vertical axis the amplitude in V and on the horizontal axis the frequency in Hz and it shows how much of the signal's energy is present at each frequency. Discrete Fourier transform

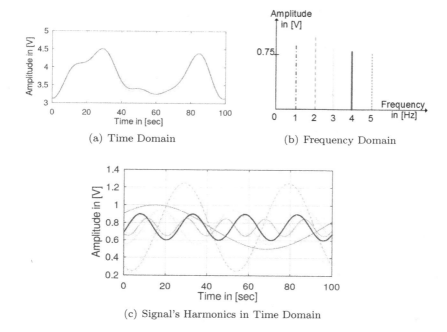

(a) Time Domain   (b) Frequency Domain

(c) Signal's Harmonics in Time Domain

FIGURE 2.5
Transformation from Time to Frequency Domain

takes digitized time domain data and computes the frequency time domain representation.

Knowing the frequencies of a signal one can define the **spectrum** of a signal as the range of frequencies that it contains. While the **bandwidth** is defined as the width of the spectrum calculated as the difference between the upper and lower frequencies within the spectrum. The bandwidth is usually measured in Hz. For example, considering the spectrum allocated for GSM 900 for uplink as 890 to 915 MHz, the bandwidth of GSM 900 uplink would be 915–890 = 25 MHz.

2.2.3 Baseband and Carrier

We define the **baseband** signal as the analog or digital signal resulted after converting it from analog or digital data. This is usually done with a frequency band (baseband) between almost zero and the maximum signal frequency, depending on the bit rate.

For example, considering the case of Public Switched Telephone Network (PSTN), the analog signals derived from voice are transmitted in the baseband 300–3400 kHz between the analog telephone and the local exchange.

The **carrier** signal is defined as a continuous constant radio frequency signal on which analog signals can be imposed by varying amplitude, frequency and phase through modulation.

For example, in GSM the analog baseband signal of 200 KHz is shifted to a carrier frequency at about 900 MHz. For wireless transmissions, it is virtually impossible to transmit baseband signals, for various reasons with some of them outlined below:

- using only baseband transmission, Frequency Division Multiplexing (FDM) could not be applied. By using analog modulation, we shift the baseband signals to different carrier frequencies. Thus, the higher the carrier frequency, the more bandwidth is available.

- the antenna size corresponds to the signal's wavelength. Consequently, considering the transmission of a babseband signal with 1 MHz bandwidth would result in an antenna some hundred meters high according to equation 2.1. Thus, this would not be useful for mobile devices.

- The characteristics of the wireless signal propagation (i.e., path-loss, penetration of obstacles, reflection, scattering, etc.) heavily depend on the signal's wavelength. Consequently, depending on the application, a carrier frequency with the desired characteristics has to be selected.

2.3 Antennas

We have seen that in wireless communication, the data is transmitted via radio signals that are represented by electromagnetic waves generated by a transmitter through modulation and emitted by the antenna of the transmitter into the air/space. The antenna at the receiver side, will catch the electromagnetic wave and demodulated by sampling the signal to recover the data bits.

Thus, an important role in wireless communication is played by the antenna which can be defined as an electrical conductor used for radiating (e.g., transmission) or collecting (e.g., reception) electromagnetic energy into/from space [1]. Thus, in the case of transmission, the antenna converts the electrical signals into electromagnetic waves. While in the case of reception, the antenna converts the energy from electromagnetic waves to an electrical signal [6]. The dimensions of an antenna depend on the wavelength, λ as defined by equation 1.1.

The radiation pattern of an antenna is a graphical representation of the field strength (voltage/meter) or the power density (watt/square meter) measured at various angular positions of constant distance to the antenna. In general, the radiation of the antenna is measured in the horizontal and in the vertical plane. The horizontal plane also known as the azimuth pattern,

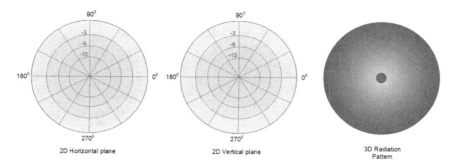

FIGURE 2.6
Isotropic Antenna Radiation Pattern

represents the radiation pattern as observed when looking at it from directly above the antenna. The vertical plane also known as the elevation pattern, represents the radiation pattern as observed when looking at it from the side of the antenna.

An **isotropic antenna** is an ideal theoretical antenna (does not exist in reality) that operates without losses and has an uniform three-dimensional radiation pattern. Figure 2.6 illustrates the perfect 360^0 vertical and horizontal beamwidth and the spherical radiation pattern of the isotropic antenna. The radiation pattern of the isotropic antenna when seen in 3D is homogeneously distributed over a spherical surface with the radiator in the centre of the spherical surface as illustrated in Figure 2.6. This antenna is used as a reference antenna when calculating the gain of real-world antennas.

Consequently, there are three fundamental properties that could be identified when dealing with antennas in wireless communications systems, such as: *gain*, *direction* and *polarization*. The **gain** is defined as the increase in energy of the radio frequency signal added by the antenna. The **direction** represents the direction of the radiation pattern of the transmitting antenna. In the case of **directional antennas**, these will be directing a beam of radiation in one or more directions referred to as **main lobes**. A *side lobe* is defined as the beam of radiation in undesired direction, and *back lobe* is defined as the beam of radiation in the opposite direction of the main lobe/beam. To provide a greater coverage distance the gain of the directional antenna can be increased at the cost of reducing the coverage angle. Thus, we define **beamwidth** in both, horizontal and vertical planes, as the angle between the points on the main lobe that are 3dB lower in gain as compared to the maximum. While, **polarization** is defined as the physical orientation of the antenna element that actually emits the RF energy [6].

The direction of the maximum radiation in the horizontal plane of the antenna is considered to be the front of the antenna, while the back is the direction 180^0 from the front. An important aspect of the antenna is the

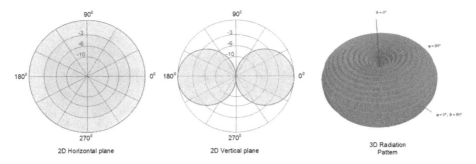

FIGURE 2.7
Dipole Antenna Radiation Pattern

front-to-back ratio which measures in fact the directivity of the antenna. The energy left behind the antenna is considered wasted. Thus, a higher gain of the antenna yields a higher front-to-back ratio.

Depending on their directionality, antennas can be classified into: *omnidirectional* and *directional* antennas. **Omnidirectional antennas** radiate in all directions in the horizontal plane. Thus, they provide a 360^0 horizontal radiation pattern while the vertical coverage varies.

In contrast to the isotropic antenna, the real commercial antennas do not radiate the same power density in all directions. An example of a real antenna, is the dipole antenna with its radiation pattern illustrated in Figure 2.7. It can be seen that the radiation pattern of the dipole antenna is omnidirectional in the horizontal plane and figure eight pattern in the vertical plane. The shape of the 3D radiation pattern resembles a donut, assuming the dipole antenna is standing vertically [7].

The gain of the dipole antenna, defined as the maximum power in the direction of the main lobe compared to the power of an isotropic radiator (with the same average power) is of 2.14 dB in the horizontal plane. Two basic examples of dipole antennas are: (1) the *half-wave dipole* or Hertz antenna, whose length is one-half of the wavelength (e.g., $\lambda/2$) of the signal that can be transmitted most efficiently and (2) the *vertical quarter-wave antenna* or Marconi antenna whose length is one quarter of the wavelength (e.g., $\lambda/4$) and is most commonly used on car roofs and portable radios [1].

2.3.1 Antennas for Mobile Devices

The antenna design for mobile devices has been a challenging task over the years. Figure 2.8 gives an illustration of the evolution of the mobile phones antennas over the years.

The mobile devices as we know them today use several radio bands such as the cellular network itself, the GPS, Wi-Fi, Bluetooth, etc. Thus, when designing antennas for mobile devices several factors need to be considered,

FIGURE 2.8
Evolution of Mobile Phones' Antennas

including antenna size, antenna shape and placement given the space restrictions of the mobile device and having multiple antennas required to be fit within an usable area [8].

We have seen that the antenna dimensions are normally multiple of a quarter or half the wavelength of the signal to be transmitted or received. However, taking for example a mobile phone operating in the GSM 900 MHz band, a half-wave dipole would need to be 333.3 mm in length. Which makes it difficult to use especially in hand portable devices. The size of the antenna can be further reduced in length by cutting it further to quarter and even one-eight wave dipoles, or by designing them in form of a helix, which has the overall correct electrical length but is packed into a smaller physical space. Additional tricks to reduce the size of the antenna could be to use the ground plane of the circuit board or zigzagging the antenna trace on the board.

With the evolution and advancements in technologies we slowly moved away from the small telescoping whip antenna located on the outside of the mobile device as illustrated in Figure 2.8 towards small board-mounted chip antennas traces printed on the circuit board. Thus, some of the most popular printed antennas are patch antennas [9–12], Inverted-F Antennas (IFA) [13–15] or Planar Inverted-F Antennas (PIFA) [16–19]. These types of antennas are preferable in contrast to the dipole antenna because they can be confined in a smaller space and they make use of the ground plane of the circuit board to radiate.

However, with the new 5G and beyond technologies new challenges are emerging for the antenna design to enable support for the new 5G frequency bands, such as sub 6 GHz bands and millimeter Wave (mmWave) frequencies above 24 GHz. The challenges for sub 6 GHz antennas involve fitting the 5G capable antenna into the compact mobile phone to work alongside existing 4G/3G and WiFi communication channels. In order to increase the data rate, support for Multiple Input Multiple Output (MIMO) multi-antenna operation needs to be considered as well, along with the increasing number of antennas for the other radios as well. The challenges for the mmWave capable antennas involve integrating the antenna inside the device behind a cover which will

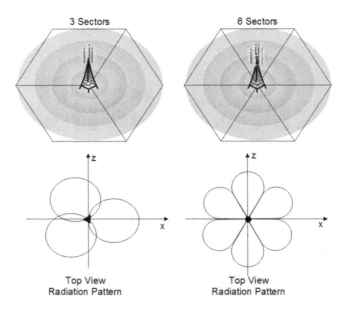

FIGURE 2.9
Sectorized Antennas for Cellular Networks

impact the radiating performance of the antenna. The size of the antennas at or above 28 GHz is relatively small, opening up new possibilities to use chip-integrated arrays [20–22]. In order to overcome the challenges of using mmWave capable antennas and improve their radiation patters, some possible solutions would be to look into the material used for mobile phone covers or to use a slot-based design integration on the rim of the phone.

2.3.2   Sectorized Antennas for Cellular Networks

Cellular networks mainly used directional antennas, strategically deployed on a single pole to construct a sectorized antenna offering 360^0 coverage. Cellular networks typically use a three sector (e.g., three 120^0 designed directional antennas) or a six sector configuration (e.g., six 60^0 designed directional antennas) as illustrated in Figure 2.9.

The ever growing demand of wireless services pushed the advancements in antenna technology while the use of multiple antennas opened up new opportunities to exploit the spacial domain. Consequently, recently the use of MIMO communications where multiple antennas are used at both transmitter and receiver, have been adopted to improve the link reliability as well as network capacity through spacial diversity and spacial multiplexing using the same time and frequency resources. The diversity gain is achieved by having multiple antennas transmitting the same signal over different paths and the signal is combined at the receiver increasing the signal quality. While multiplexing gain is achieved by having multiple antennas transmitting multiple signals over different paths increasing in this way the data rate [8].

The existing 4G networks have already integrated the deployment of 2x2 MIMO with increasing progress towards the adoption of 4x4, 8x2 or even 8x8. This trend is not an issue for the base station at the cellular network side as they are not resource constrained. However, the limitation appears at mobile device side because of the form factor and energy efficiency. Most of the latest smartphones can support 4x4 MIMO configurations already which enables an increase in data rates.

The evolution in antenna technology resulted in a variety of MIMO applications, such as: (1) Single User-MIMO (SU-MIMO) [23] which is in fact the conventional MIMO case with one base station and one mobile device equipped with MIMO antennas; (2) Multi User-MIMO (MU-MIMO) [24] where a MIMO base station can transmit multiple signals to multiple users simultaneously equipped with single or multiple antennas; (3) Full-Dimension MIMO (FD-MIMO) [25] for penetrating into buildings and skyscrapers more effectively and (4) massive MIMO (mMIMO) which uses increased number of advanced multi-band antennas.

5G sees the integration of mMIMO solutions with configurations of 16×16, 32×32 and 64×64, and potentially higher combinations in the long-term, including larger number of antennas for both passive and Advanced Antenna System (AAS). This is possible due to the use of the higher frequency bands which means a reduce in size for individual antenna elements enabling the use of more antenna elements within the same physical space. The concept of **beamforming** refers to the ability of an AAS to direct (transmission)/to collect (reception) radio energy through/from the radio channel towards/from a specific receiver/transmitter.

The evolution from passive to AAS is illustrated in Figure 2.10. A conventional base station consists of three main components: a Digital Unit (DU) or Baseband Unit (BBU) that handles the processing of the digital signal, radio unit for generating the analog RF signal and the passive antenna unit located in the mast, to emit the RF signals. In Generation 1 as illustrated in Figure 2.10, the radio unit and the DU are deployed underneath, at the base of the site, within an Base Transceiver Station (BTS) unit connecting to the antenna through a long RF feeder cable, which results in substantial power losses [23].

Slowly, as the radio hardware reduced in size, the deployment of Remote Radio Head (RRH) closer to the antenna became common starting with Generation 2. In this case a long optical fiber is used to connect the DU to the RRH while a short RF feeder cable connects the RRH to the antenna. One single DU can be used to support several antennas deployed at the same site, that could cover different sectors or use different frequency bands. The next Generations moved from passive antennas to active antennas that integrated the antenna and the RRH into the same unit. Different configurations of active antenna systems are possible, from single antenna units connected to form a

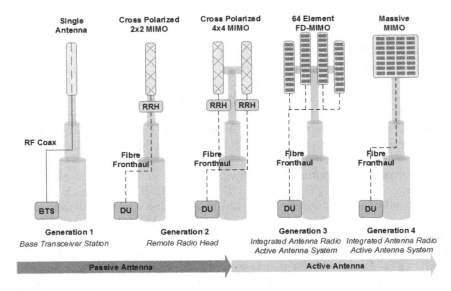

FIGURE 2.10
Passive vs. Active Antennas

FD-MIMO system to intelligent mMIMO systems that can use beamforming to adapt the radiation patterns [26]. These active mMIMO antenna arrays systems are known within industry as AAS and they are becoming a dominant approach for 5G networks due to their benefits of enhancing the communication throughput, reduce cable losses, reduced cost and power consumption [24]. Additionally, the configuration and deployment of AAS is flexible, in the case of small cells even the DU can be integrated within the antenna unit while in the case of macro cells, one DU can be shared between multiple active antennas or can be moved to the cloud to form the Cloud Radio Access Network (C-RAN) [27].

2.4 Multiplexing and Modulation

Figure 2.11 illustrates a typical wireless transmission chain block diagram. It includes a transmitting station and the receiving station communicating over the air interface. The stations consist of the source encoding/decoding blocks, channel encoding/decoding, encryption/decryption, multiplexing/demultiplexing, modulation/demodulation. All these signal processing blocks can be optimized individually to enable reliable communication between the transmitter and receiver over the air interface [28].

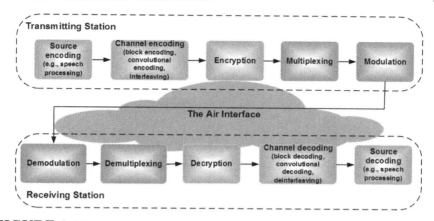

FIGURE 2.11
Wireless Transmission Chain – Overview

2.4.1 Multiplexing Techniques

The multiplexing techniques are used to allow more than one user (e.g., mobile devices) to share a common communication medium, such as the air interface. Sharing the communication medium requires an access mechanism (see Section 2.6) that will control the multiplexing mechanism [28].

Multiplexing for wireless communications can be done in four dimensions, such as: space (s), time (t), frequency (f) and code (c) [3]. The aim is to achieve a maximum utilization of the medium with minimum interference, by assigning space, time, frequency or code to each communication channel represented here by the association of the senders and receivers.

Figure 2.12 illustrates five communication channels (k_i) and different types of multiplexing techniques represented on the three dimensional coordinate system given by time (t), frequency (f) and code (c). The **Space Division Multiplexing (SDM)** makes use of the space s_i represented by the circles indicating interference range and the mapping of each channel onto a dedicated space (e.g., channels k_1, k_2 and k_3 mapped into spaces s_1, s_2 and s_3). **Guard spaces** are used to space out the channels and enable non-overlapping interference ranges between them. These guard spaces are used in all multiplexing techniques to avoid interference. In wireless communications SDM is enabled by sectorized/directional antennas and it is usually used in combination with time, frequency or code division multiplexing techniques [6].

In the case of **Time Division Multiplexing (TDM)**, the time is divided into time slots and one channel has the whole spectrum for a certain amount of time (e.g., time slot). Thus, all users are going to use the same frequency and code but at different time slots. The time slots are separated by guard spaces that represent time gaps in order to avoid co-channel interference, where two channels overlap in time. The advantage of TDM is that it can

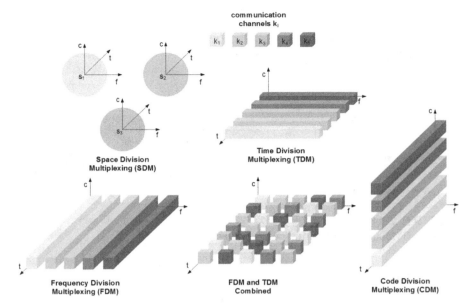

FIGURE 2.12
Multiplexing Techniques

achieve high throughput as it is using the whole bandwidth and multiple time slots can be allocated to users that have more data to send. However, the main disadvantage is that all senders need to be precisely synchronized to avoid overlapping in time while the receivers need to tune in at the exact point in time as well to be able to communicate with the senders.

Frequency Division Multiplexing (FDM) involves the division of the whole spectrum into smaller frequency bands as illustrated in Figure 2.12. In this case, a channel will be allocated a certain band of the spectrum for the whole duration/time. To avoid overlapping channels which could lead to adjacent channel interference, the guard spaces are used representing frequency gaps. The main advantage of FDM is that there is no need for dynamic coordination between the senders and receivers as the receiver just needs to tune into the specific frequency of the sender. However, due to the use of guard spaces there is a good amount of bandwidth wasted, which can also happen if the traffic is distributed unevenly. There might be situations where the channel is allocated even if there is no data to send. This permanent assignment of a frequency to a sender makes the FDM approach inflexible and inapplicable for mobile communications where the assignment of a permanent frequency band for each mobile device would be firstly impossible and secondly, a tremendous waste of scarce frequency resources. However, FDM is suitable for radio broadcast transmissions (e.g., 24 hours a day) [29].

An interesting solution is to combine the use of FDM and TDM as illustrated in Figure 2.12. In this context, a channel would use a certain frequency band for a certain time slot. Guard spaces are used in both time and frequency dimensions, to avoid interference. The main advantage of this technique is that it provides a better protection against *frequency selective interference* where there is interference in a certain small frequency band. This is because, by combining FDM and TDM, the channel will not use this small frequency band for the entire duration but only for a time slot before jumping to another small frequency band that might not have interference. However, in order for a sender receiver pair to be able to communicate, they will need to coordinate on the exact sequence of frequencies and corresponding time slots. This brings the advantage that it provides increased protection against tapping, as it requires the full sequence of frequencies to be known to be able to listen into the channel. However, the main disadvantage of this technique is the precise coordination between senders as well as between sender and receiver where the exact mapping of the frequency sequence to time slots is required.

In **Code Division Multiplexing (CDM)**, all channels use the same frequency band at the same time, but the separation is done by the use of different codes. Guard spaces are used to provide a code gap and avoid interference. Ideally these codes need to be orthogonal enough to separate the communication channels. A sender and a receiver need to share the same code to be able to communicate. Thus, this method offers a good protection against interference and tapping and because of this it was initially used in military applications. Because all the users are using the same frequency at the same time, the receivers must be able to separate the user data from the background noise consisting of other users' signals as well as environmental noise. Thus, the high complexity at the receivers does not make this method popular.

A special method of data transmission is **Orthogonal Frequency Division Multiplexing (OFDM)**. OFDM builds on simpler FDM where the whole spectrum is divided into several spaced out frequency sub-channels to avoid overlapping or interference. However, OFDM makes use of orthogonal overlapping frequencies sub-channels that are carefully chosen and modulated so that the interference between them is canceled out as illustrated in Figure 2.13. It can be noted that the peak at the center carrier frequency of each sub-carrier corresponds to a zero level of every other sub-carrier. Thus, there is no adjacent channel interference. If the receiver samples at the center frequency of each sub-carrier, the only energy present is that of the desired signal. Consequently, OFDM is more bandwidth efficient than conventional FDM as the guard bands between the sub-carriers are eliminated and they can be spaced out narrowly [29].

Moreover, OFDM forms the basis for *multi-carrier modulation* where a single stream of digital information is split among these orthogonal narrowband subchannel frequencies instead of using one single wideband channel frequency. Thus, this allows for several bits to be sent in parallel or at the same time but in separate channels.

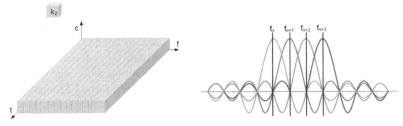

Orthogonal Frequency Division Multiplexing (OFDM)

FIGURE 2.13
Orthogonal Frequency Division Multiplexing

2.4.2   Modulation Techniques

The next block after *multiplexing* in the typical wireless transmission chain diagram in Figure 2.11 is *modulation*. **Modulation** represents the modification of a carrier signal parameters: amplitude, frequency, phase or a combination of them, in dependence of the symbol that needs to be sent. The *symbol* represents an abstract quantity that carries a single bit or several bits at once and that is assigned to a certain signal state. The *signal state* represents one of several constellations of a carrier's parameters defined by the used modulation scheme. For example, if a symbol can take m possible values, the amount of bits carried by the symbol is given by $n = \log_2 m$ bits/symbol. The *symbol duration* is given by time T_s the signal state is kept by the transmitter for transferring the symbol. The *symbol rate* represents the number of transmitted symbols per second given by $R_s = 1/T_s$ measured in symbols per second. The *data rate* is defined as $R_d = n \cdot R_s$. The *chip* represents a very small pulse of -1 or $+1$ of duration T_c, with $T_c << T_s$. The *pulse* is a representation of bits or symbols, usually in the baseband. For example, using a bipolar or polar notation.

Digital transmissions cannot be used in wireless networks. Thus, digital data has to be transmitted over a medium that only allows for analog transmission [3]. Consequently, digital data has to be first translated into analog signals through digital modulation while analog modulation is then used to shift the center frequency of the baseband signal up to the radio carrier where the signal will be ready for transmission via the antenna. Three basic digital modulation schemes are Amplitude Shift Keying (ASK), Frequency Shift Keying (FSK) and Phase Shift Keying (PSK). While three basic analog modulation schemes are Amplitude Modulation (AM), Frequency Modulation (FM) and Phase Modulation (PM) [28].

One of the most simplest digital modulation schemes is **Amplitude Shift Keying (ASK)** as illustrated in Figure 2.14. Given a set of digital data consisting of 0s and 1s and represented by the corresponding pulses, where a

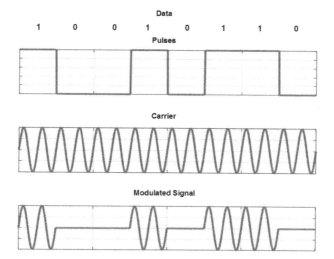

FIGURE 2.14
Amplitude Shift Keying – Example

binary 0 can be encoded as zero or negative voltage, and a binary 1 is encoded as a positive voltage. ASK represents the two binary values of 0s and 1s using two different amplitudes of the carrier frequency as indicated in equation 2.4. In this example, the binary 0s is represented by amplitude zero.

$$s(t) = \begin{cases} A\cos(2\pi f_c t), & \text{binary 1} \\ 0, & \text{binary 0} \end{cases} \qquad (2.4)$$

The main advantages of ASK is that it is very simple and requires low bandwidth. However, very susceptible to interference due to small-scale fading, which causes rapid fluctuations of the signal's amplitude [3]. Consequently, it is not used for wireless radio transmissions (apart from infrared systems) because a constant amplitude value cannot be guaranteed. However, this method is favoured for optical transmissions in wired networks.

Another simple digital modulation scheme that is commonly used for wireless transmissions is **Frequency Shift Keying (FSK)** as illustrated in Figure 2.15. In this case, the two binary values of 0s and 1s are represented by two different frequencies, frequency f_1 and frequency f_2 as given by equation 2.5.

$$s(t) = \begin{cases} A\cos(2\pi f_1 t), & \text{binary 1} \\ A\cos(2\pi f_2 t), & \text{binary 0} \end{cases} \qquad (2.5)$$

There are two types of FSK schemes: (1) *Continuous phase modulation* where the frequency changes occur at the carrier zero crossings to avoid sudden changes in phase which could lead to high frequencies and (2) *Non-continuous*

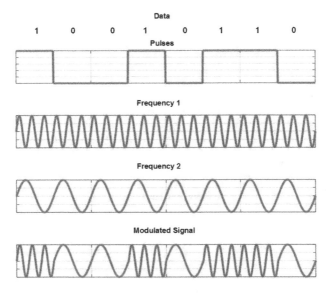

FIGURE 2.15
Frequency Shift Keying – Example

phase modulation where the frequency changes occur between the carrier zero crossings which results in increased spectral spreading and thus increased out-of-band radiation.

A special form of FSK that is often used in many wireless systems is **Minimum Shift Keying (MSK)** which provides a continuous phase and keeps synchronization between the two frequencies f_1 and f_2. MSK works by separating the bits from the digital information data into *even* and *odd* (based on their position) and doubling the bit duration as illustrating in Figure 2.16 [3]. Thus, the even bits are formed by bringing down the bits on the even positions within the data and doubling their duration. Similarly, the odd bits are formed by bringing down the bits located on the odd positions within the data and doubling their duration. In this example, the first bit of the even bits pulses was assumed to be a negative voltage level.

Once the bits separation is completed the MSK modulated signal is formed based on equation 2.6.

$$s(t) = \begin{cases} \text{invert} f_2, & \text{if even bit} = 0 \ \& \ \text{odd bit} = 0 \\ \text{invert} f_1, & \text{if even bit} = 1 \ \& \ \text{odd bit} = 0 \\ f_1, & \text{if even bit} = 0 \ \& \ \text{odd bit} = 1 \\ f_2, & \text{if even bit} = 1 \ \& \ \text{odd bit} = 1 \end{cases} \tag{2.6}$$

A higher bandwidth efficiency is achieve by MSK when using a Gaussian low-pass filter that will reduce even further the MSK sidelobe levels. This new

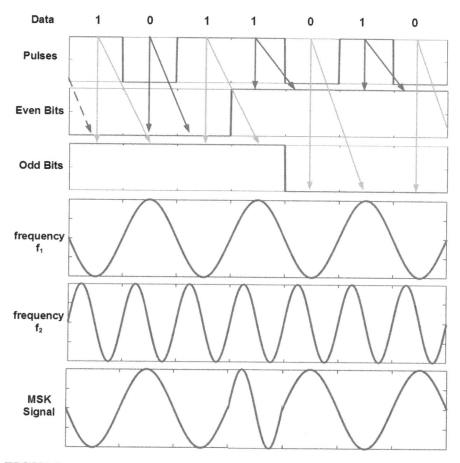

FIGURE 2.16
Minimum Shift Keying – Example

modulation scheme obtained is referred to as **Gaussian Minimum Shift Keying (GMSK)** and is used in GSM.

The third simplest modulation scheme **Phase Shift Keying (PSK)** represents the data by making use of the shifts in the phase of the signal as illustrated in Figure 2.17. The simplest version of PSK, also known as **Binary Phase Shift Keying (BPSK)**, makes use of two phases to represent the two binary digits of 0s and 1s as given in equation 2.7. Consequently, a shift of 180^0 happens each time there is a transition in the digital data from 0 to 1 or vice versa.

$$s(t) = \begin{cases} A\cos(2\pi f_c t), & \text{binary 1} \\ A\cos(2\pi f_c t + \pi), & \text{binary 0} \end{cases} \tag{2.7}$$

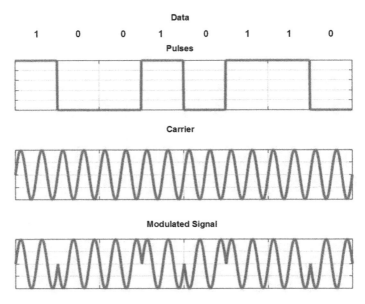

FIGURE 2.17
Phase Shift Keying – Example

In a differential PSK version of the modulation scheme, a binary 0 would be represented by sending a signal burst of the same phase as the previous signal burst and a binary 1 would be represented by sending a signal burst of opposite phase to the preceding one.

The modulation schemes, we have seen until now, define two signal states for representing binary 0 or 1 (e.g., this is given by 2^1 because we transfer one bit only, of 0 or 1, in a signal modulation step). However, in theory a digital modulation scheme could actually fix an arbitrary number of signal states which could help at increasing the bits transferred in a signal modulation step. Thus, one option is to improve PSK by using multiple signal states. For example, Quadrature Phase Shift Keying (QPSK) uses four signal states to transmit two bits in one signal modulation step as illustrated in Figure 2.18.

Figure 2.18 also represents the signal using the phase domain where the X-axis is called *In-Phase (I)* and Y-axis is called *Quadrature-Phase (Q)*. QPSK is used in Universal Mobile Telecommunications Service (UMTS) and IEEE 802.11 networks. The advantage of QPSK is that it requires less bandwidth as compared to BPSK, however it is more complex.

More advanced modulation schemes could be constructed by mixing and combining different types of shift keying schemes. One such example is Quadrature Amplitude Modulation (QAM) which combines ASK and PSK.

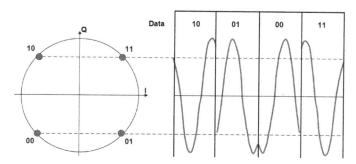

FIGURE 2.18
Quadrature Phase Shift Keying – Example

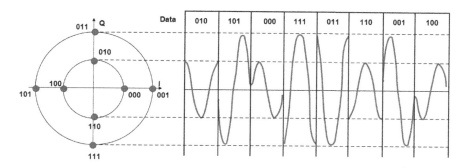

FIGURE 2.19
Quadrature Amplitude Modulation – Example

In this way, QAM uses eight signal states (2^3) to transmit three bits in one signal modulation step as illustrated in Figure 2.19.

Nowadays, due to the advancements in technology it is possible to use a higher order modulation scheme, for example 256QAM which can transmit eight bits in one single modulation step (e.g., symbol). However, the higher order modulation used the greater the susceptibility to noise and interference. Thus, the Bit Error Rate (BER) increases with the number of bits per symbol [6].

The performance of any modulation scheme can be evaluated by looking at their *bandwidth efficiency* and *power efficiency*. The *bandwidth efficiency* refers to the ability of a modulation scheme to accommodate data within a limited bandwidth. Bandwidth efficiency is defined as the ratio of the data rate per Hertz indicating how efficiently the allocated bandwidth is utilized. One can increase the data rate by decreasing the pulse width of the digital symbol. However, this would increase the bandwidth of the signal.

On the other side, *power efficiency* refers to the ability of the modulation scheme to preserve the fidelity of the digital message at low power levels in

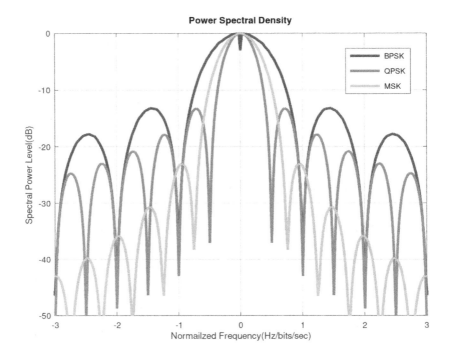

FIGURE 2.20
Power Spectral Density – Comparison BPSK, QPSK, MSK

terms of bit error probability. Thus, the power efficiency of any modulation scheme reflects how favourably the trade-off between fidelity and signal power is made. The increase in signal power is required when trying to obtain a certain level of bit error probability and to increase the immunity to noise of any modulation scheme.

An easy way to evaluate the modulation scheme in terms of *bandwidth efficiency* and *power efficiency* is by plotting their *power spectra* as illustrated in Figure 2.20 comparing BPSK, QPSK and MSK. The **Power Spectra Density (PSD)** represents the power per unit of frequency and displays the relative power contribution of the various frequency components [30]. In Figure 2.20 the width of the main lobes indicate the bandwidth efficiency while the peak of the side lobes indicate the power efficiency and out-of-band radiation. The **out-of-band radiation** represents the amount of transmitted signal power that lies outside the main lobe which leads to *Adjacent Channel Interference (ACI)*. ACI appears when a transmitting radio interferes with the frequency channels located immediately above and below of the transmitting user's frequency channel. One way to reduce ACI is to make use of filters.

In Figure 2.20 we can notice that MSK has comparatively low side lobes and moderate width of the main lobe when compared to BPSK and QPSK.

This makes MSK a favourite modulation scheme for the systems where users are separated by frequency channels, like GSM or Bluetooth. On the other side, QPSK has the lowest width of the main lobe making it bandwidth efficient. However, its side lobs are quite strong when compared to MSK

A common impairment in wireless communications that is also considered one of the main constraints on achieving bandwidth efficiency is *noise*. Thus, for a given limited bandwidth we would like to achieve the highest data rate possible within a particular limit of error rate [1].

Assuming a noiseless channel, **Nyquist** theorem states that if a bandwidth limited signal is sampled at regular intervals of time and at a rate equal or higher than twice the highest significant frequency, the sample contains all the information of the original signal. Thus, this means that given a bandwidth B Hz, the maximum data rate that can be accommodated for a signal with two discrete signal states/elements or voltage levels is $2B$ bps. Thus, we can define the Nyquist formula as given in equation 2.8 [4].

$$C = 2B \log_2 V \tag{2.8}$$

Where C is the channel capacity (bps), B is the bandwidth (Hz), V represents the discrete signal states/elements or voltage levels. We can use signals with more than two levels. For example, if we want to transmit three bits, our signal will consist of $2^3 = 8$ voltage levels.

It is important to note that Nyquist theorem assumes noiseless channels. Thus, considering for example a noiseless 3 kHz channel, it cannot transmit binary two-level signals at a rate exceeding 6000 bps.

However, in reality it is not possible to reach the performance of Nyquist's theorem due to noise that is always present in the channel. Otherwise, modulation schemes with a very high number of signal states V could be used to sent a high number of bits within small bandwidth.

Consequently, **Shannon** theorem defines the maximum data rate in bps of a noisy channel of bandwidth B Hz and the Signal to Noise Ratio (SNR) as given in equation 2.9 [4].

$$C = B \log_2(1 + SNR) \tag{2.9}$$

It can be noticed that the Shannon formula is not dependent on the number of signal levels, which means that regardless of the amount of signal levels we have, we cannot achieve a data rate higher than the capacity of the channel [5]. For example, considering a voice-band channel of 3000 Hz bandwidth, and a SNR of 30 dB, the maximum data rate the channel can transmit is of 30,000 bps, regardless of the number of signal levels used.

Signal to Noise Ratio (SNR) expresses in decibels the amount by which the signal level exceeds the noise level in a specified bandwidth as given in equation 2.10 [1].

$$SNR[dB] = 10 \log_{10} SNR \tag{2.10}$$

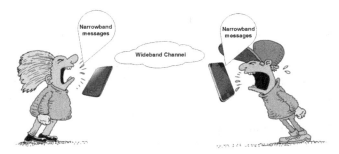

FIGURE 2.21
Spread Spectrum Example

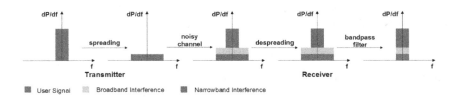

FIGURE 2.22
Spread Spectrum Process

Where SNR is defined as the average signal power divided by the average noise power.

2.5 Spread Spectrum

We have seen until now that the goal of the modulation schemes is to achieve a greater power and bandwidth efficiency. **Spread spectrum** is designed to be used in wireless applications and the spread spectrum techniques employ a transmission bandwidth that is several orders of magnitude greater than the minimum required signal bandwidth as illustrated in Figure 2.21. Thus, in the case of single user scenario, the spread spectrum is bandwidth inefficient. However, in the case of multi user environments, spread spectrum become bandwidth efficient as many users could simultaneously use the full spectrum without significantly interfering with each other.

Figure 2.22 illustrates the process involved in the spread spectrum technique when transmitting a user signal to a receiver, represented in terms of power density dP/df versus frequency f. At the transmitter side the narrowband frequency range is deliberately varied through spreading, resulting in a broadband signal. It can be noted that the power level of the spread signal can be much lower than the original narrowband signal without losing data [3]. The spreading is done according to a spreading sequence. The resulted broadband signal is sent over the noisy channel where narrowband and broadband interference will add to the signal. The receiver will receive the noisy signal consisting of the user data broadband signal plus the narrowband and broadband interference signals. Upon reception the receiver will perform despreading of the signal using the same spreading sequence. This involves converting the user data broadband signal back to narrowband signal, while the narrowband interference is spread out. Additionally, a bandpass filter could be used at the receiver to cut off the frequencies on the sides of the narrowband signal.

There are two well-known spread spectrum techniques: Frequency Hopping Spread Spectrum (FHSS) and Direct Sequence Spread Spectrum (DSSS).

When using the **Frequency Hopping Spread Spectrum (FHSS)** technique, the signal is spread **during** modulation. FHSS uses a number of different carrier frequencies that are modulated by the source signal. The carrier frequency is rapidly changed according to a spreading sequence/hopping sequence. The same spreading sequence is available at the receiver as well so that both transmitter and receiver hop on the same channel. FHSS is based on a combination of FDM and TDM as the transmitter and receiver stay on a certain channel for a certain amount of time only before they hop to another channel. FHSS brings several benefits, such as: (1) certain frequency channels may be temporarily interfered (e.g., adjacent channel, co-channel, inter-symbol interference). However, by using FHSS only a few bits might be destroyed at the interfered frequency channel instead of the entire transmission. (2) the lost bits have only a minor impact on the overall transmission (e.g., voice transmission) or a re-transmission mechanism could be employed such as Automatic Repeat Request (ARQ) in case of data transmission.

On the other side, in the case of **Direct Sequence Spread Spectrum (DSSS)** the signal is spread **before** modulating it onto a carrier signal with fixed carrier frequency. Thus, the data signal, rather than being transmitted on a narrowband, is spread onto a much larger range of frequencies using a specific spreading sequence (e.g., pseudo-noise sequence). This is achieved by performing an XOR operation between the user data and the spreading sequence, which is a chipping sequence in this case consisting of smaller pulses called chips. If generated properly, this chipping sequence will appear as random noise, e.g., pseudo-noise sequence. Within a wireless environment, the impact of multipath propagation can cause attenuated, phase-shifted and time-delayed copies of a transmitted signal. In this context FDM/TDM is not an efficient approach to support variable data rates. Thus, when using DSSS it brings several benefits: (1) DSSS signal is very robust as it has a redundancy factor; (2) the effects of multipath propagation can easily be detected when using spreading codes with good auto-correlation; (3) data rate depends on the length of the spreading code [6].

We note that a spreading sequence should have a good correlation properties, which determines how much similarity one set of data has with another set. When analyzing the correlation of spreading sequences we can look at *auto-correlation* and *cross-correlation* properties. It is said that a spreading sequence has a good *auto-correlation* if the inner product with itself is large and its inner product with the same, but shifted spreading sequence is small. Having a good auto-correlation is essential as it could improve the overall system performance through good synchronization between sender and receiver and good anti-jam performance. An example of a spreading sequence with goo auto-correlation, that is used in IEEE 802.11 is the Barker code (10110111000). When using the Barker code with DSSS, the user signal with a bandwidth of 1 MHz is spread (user data XORed with the spreading sequence) into a signal with 11 MHz bandwidth (note 11bit Barker code). After the signal is spread it is then modulated onto a carrier frequency (e.g., 2.4 GHz in the ISM band for IEEE 802.11) before being transmitted.

On the other side, it is said that two spreading sequences have a *low cross-correlation* if their product is low for all shift combinations. If there is no relation at all between the two spreading sequences then the cross-correlation is zero and they are said to be *orthogonal*. Having low cross-correlation properties for the spreading sequences is useful to filter out noise, and the receiver could discriminate among signals generated by different users.

FHSS provides a simpler spread spectrum technique than DSSS. Compared to DSSS, the FHSS solution uses only a small portion of the total bandwidth available at any time while DSSS will always use the full bandwidth available. However, with DSSS the system is more resistant to fading and multi-path effects.

Consequently, it can be noted that spread spectrum systems bring many advantages: (1) being initially invented for military purposes, the spread spectrum technique is very secure as the user signal is very hard to distinguish from the background noise and the interceptor needs to know the spreading sequence to be able to decode the data; (2) spread spectrum systems are very robust, as the user signals can be received even in the presence of very strong narrowband interference, and they also have a better performance in the presence of multi-path fading; (3) the spread spectrum systems enable new transmission technologies to be overlaid at the same frequency; (4) can be used as the basis for a further multiplexing/multiple-access technology.

2.6 Medium Access Mechanisms

The data link layer of the OSI reference model [31] consists of two sub-layers: Logical Link Control (LLC) and Medium Access Control (MAC). The MAC sublayer is responsible for the medium access resolution within a multi user environment. The main role of the access methods for wireless communication systems is to orchestrate the successful operation of multiple terminals over the wireless medium using SDM, TDM, FDM or CDM. However, most of the access methods have been originally developed for wired communication. Thus, in order to make them suitable for wireless communications, they require some changes/updates to optimize their performance to the specifics of the unreliable wireless medium. We can classify the existing medium access solutions into two wide categories, such as: fixed-assignment access and random access solutions.

2.6.1   Fixed-Assignment Access Solutions

The fixed-assignment originates from the telecommunications domain where the access methods assign a slot in time, a portion of frequency or a specific code to the user, preferably for the entire duration of the conversation.

Frequency Division Multiple Access (FDMA) is built upon FDM where the available bandwidth is divided into frequency channels and each user is allocated one channel for the entire duration of the call. Thus, using FDMA all users can transmit signals simultaneously, and they are separated from one another by their frequency of operation. The frequency channel allocation is done on demand, usually by the Base Station (BS) based on the incoming requests from the users. The frequency channels are separated by guard spaces to avoid interference. The advantage of FDMA is that it requires low complexity as compared to Time Division Multiple Access (TDMA) and Code Division Multiple Access (CDMA) and it needs less synchronization and framing overhead [5].

The BS and the mobile terminal transmit simultaneously and continuously by using **Frequency Division Duplex (FDD)**. In this context, each user is assigned two frequencies: one for downlink or forward channel indicating the direction of communication from the BS to the mobile terminal and one for uplink or the revers channel indicating the direction from the mobile terminal to the BS.

For example, FDMA and FDD are used in GSM 900 where the band between 890.2 and 915 MHz is used for uplink and the band between 935.2 and 960 MHz is used for downlink. The frequencies for uplink and downlink channels are allocated by the BS to the mobile terminal.

Time Division Multiple Access (TDMA) is built upon TDM where a single frequency carrier is shared by many users. In this context, time slots are introduced and in each time slot only one user can transmit or receive using the whole spectrum. Guard space in time are introduced to avoid overlapping time slots which could lead to interference. A frame consists of a N number of time slots and a channel is defined as a particular time slot that reoccurs every frame. This means that the transmission for any user is non-continuous. However, the advantage of using TDMA is that it is possible to enable the support for users with varying data dates by allocating different number of time slots per frame for different users. The number of time slots depends on the modulation technique, available bandwidth, desired bit rate, etc [8]. However, the main disadvantage of this solution is that it requires synchronization between mobile terminals and the need for guard spaces.

The BS and the mobile terminal can also use **Time Division Duplex (TDD)** for communication. TDD uses time instead of frequency to provide forward and reverse communication channels. If the time separation between these channels' time slots is small, then the transmission and reception of data appears simultaneous. However, due to the time latency created by TDD, it is not a full-duplex in the truest sense.

In general, the communication between two devices can be: (1) *simplex* – where the communication is happening unidirectional, only one device can transmit and the other can receive only; (2) *half-duplex* where both devices can transmit and receive, but not at the same time, and (3) *full-duplex* where both devices can transmit and receive simultaneously [8].

For example, TDMA is used in Digital Enhanced Cordless Telecommunications (DECT) where there ten carrier frequencies with 24 time slots per frame each, 12 time slots allocated for downlink and 12 time lost for uplink. The time slot has a duration of 417 μs and the frame lasts for 10 ms. Moreover, apart from FDMA GSM is also using TDMA, thus it combines the two access mechanism and for each carrier frequency a TDMA frame consisting of eight time slots is used. The duration of a time slot is of 576.9 μs making the communication appear simultaneous. The channel is represented by a time slot within the TDMA frame on a certain carrier frequency given by the FDMA.

Code Division Multiple Access (CDMA) is built upon CDM where the users share the same frequency at the same time, but they are separated by code. An analogy could be the scenario of an international conference where people from different countries speaking different languages are located in the same large room. In this context, the people are the users and the common medium they are sharing is the space in the large room. Any two users to be able to communicate they need to speak the same language (code). Thus, everyone will speak at the same time, but only the people that share the same language are able to understand, the rest will represent background noise for any given communication. For example, you can always have two persons speaking Chinese able to understand each other while other people speaking English will be representing the background noise. Thus, by using CDMA the transmitter and the receiver need to share the same code to be able to communicate. The two spread spectrum technologies DSSS and FHSS are based on CDMA where the code is represented by the spreading sequence.

In CDMA-based systems it is possible for the user with the strongest received signal strength to successfully capture the intended receiver, even when many other users are also transmitting, as all users are transmitting simultaneously at the same time and on the same frequency carrier. Often,

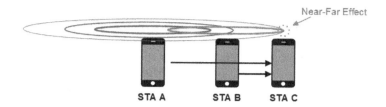

FIGURE 2.23
Near-Far Effect – Example

the closest transmitter is able to capture a receiver because of the small propagation path loss. However, a common problem that appears in this case is
the so called **near-far effect** as illustrated in Figure 2.23 where two stations
(e.g., STA A and STA B) are transmitting to a third one (e.g., STA C). However, STA A is located far from STA C while STA B is located near STA C.
Consequently, the signal of STA B drowns out STA A' signal due to large
path loss of STA A's signal. To be able to resolve this problem, a power control mechanism is required such that STA A and STA B could adapt their
transmission power (e.g., increase or decrease) so that STA C would receive
their communication at the same power level.

Orthogonal Frequency Division Multiple Access (OFDMA) is
built upon OFDM where orthogonal frequency carriers are used and several
user devices can transmit at the same time over their allocated sub-set of
frequency carriers. For example, the Long-Term Evolution (LTE) cellular systems is using OFDMA as an access technique to accommodate multiple users
in a given bandwidth.

Space Division Multiple Access (SDMA) is built upon SDM which
enables the sharing of wireless channels among users through physical separation. For example, in cellular systems, the directional antennas are used to
provide connectivity over separated cell sectors. Also the cell sites are spaced
out to avoid interference. However, SDMA needs to be used in conjunction
with the other access methods to be efficient.

2.6.2   Random Access Solutions

The random access solutions originate from data communications and provide
a more flexible and efficient way of managing channel access. These methods
provide the users with varying degrees of freedom in gaining access to the
medium whenever there is information that needs to be sent. However, the
randomness nature of the user access to the medium may cause contention
resulting in collisions of contenting transmissions [5].

One of the most simplest and popular medium access protocol that has
been developed at University of Hawaii, is **Pure Aloha** [32]. The aim of this

protocol was to provide a simple way to interconnect computers on the island campuses with the University's main computer center on Oahu and it was put in operation in 1971. The principle behind *Pure Aloha* is that a station can access the medium as soon as there is a message that is ready to be transmitted. Consequently, if two or more stations access the medium at the same time, a collision occurs and transmitted data is lost. An acknowledgement is expected after each transmitted message. If a collision occurs, the transmitting station will wait for a random amount of time and will re-transmit the message. As the number of devices accessing the shared medium increases, a greater delay occurs which is due to the increase in the probability of collisions.

To improve the efficiency of the system from 18% up to 37%, **Slotted Aloha** was introduced. The principle behind *Slotted Aloha* is that the transmission time is divided into time slots and when a station has data ready for transmission, the data packet is buffered and transmitted at the start of the next time slot. Thus, the advantage is that the partial packet collision is eliminated as compared to Pure Aloha, in Slotted Aloha there is either no collision or a complete collision. However, all stations need to synchronize their time slot to a beacon signal emitted by a special station.

Pure Aloha and Slotted Aloha work well under light traffic load. However, their performance degrades significantly under heavy traffic. Consequently, to overcome these issues reservations are introduced that could increase the system efficiency even further up to 80%. **Reservation ALOHA** or Demand Assigned Multiple Access (DAMA) is an explicit reservation scheme where the stations have to reserve explicitly each transmission slot. The time slots are divided into periods of reservations and periods of transmission. During the reservation period, the stations can reserve future slots to use for data transmission within the transmission period. The stations access the reservation period according to Slotted Aloha. Thus, collisions can only occur during the reservation period.

Packet Reservation Multiple Access (PRMA) is an implicit reservation scheme where a certain number of time slots form a frame and frames are repeated over time. Similarly to DAMA, stations can compete for empty slots according to the Slotted Aloha principle. Once the reservation was successful, the time slot is automatically assigned to requested station in all the following frames as long as the station has data to transmit. As soon as the slot was empty in the last frame, the competition for its reservation starts again.

We have seen until now, that the solutions based on Aloha do not listen to the medium before the transmission. However, great improvements in system efficiency can be achieved by listening to the communication medium before a transmission takes place. In this regard, **Carrier Sense Multiple Access (CSMA)** works by having each station in the network monitoring the status of the communication medium before transmitting the data. If the medium is idle, the user is allowed to transmit a packet based on a particular algorithm common to all transmitters in the network. Thus, the basic concept of CSMA is exposed in the following steps:

- considering two stations STA A and STA B having information to transmit

- in the first instance STA A senses the medium being idle first and then sends a packet

- STA A senses the idle medium and transmits again

- During the second transmission of sender A, sender B senses the medium and discovers that the medium is in use

- Sender B delays its transmission using a back-off algorithm and collisions are avoided

However, even in CSMA due to long propagation delays, sensing fails and transmissions may overlap which leads to collision and packet loss. There are two types of CSMA methods, such as: non-persistent and persistent. There are two non-persistent CSMA methods: (1) *Unslotted Non-persistent CSMA* where after sensing the medium busy, the stations will attempt another sensing only after a random waiting period and the transmission starts as soon as the medium is sensed free. (2) *Slotted Non-persistent CSMA* where after sensing the medium busy, the station attempts another sensing after a random number of time slots and the transmission starts at the beginning of a time slot.

In the case of persistent CSMA methods, there are also two methods: (1) *unslotted p-persistent CSMA* where station senses the medium continuously until it is free and it starts the transmission with a probability p or defers the transmission by half of the Round Trip Delay (RTD) with probability $1 - p$ before sensing the medium again. (2) *slotted p-persistent CSMA* where the station senses the medium at the beginning of each time slot until the medium is sensed free. Once the medium is free, the station starts the transmission with probability p or defers the transmission to the next time slot with probability $1 - p$ before sensing the medium again.

However, there are two particular issues related to wireless networks that the CSMA methods cannot handle, such as hidden terminals or exposed terminals.

Figure 2.24(a) illustrates the **hidden terminal problem**. Two terminal STA A and STA C can each be within the range of some intended third terminal STA B, but out of the range of each other. This is because STA A and STA C are separated by excessive distance or by some physical obstacles that makes their direct communication impossible. Assuming that STA A is transmitting to STA B and that STA C also wants to transmit to STA B and senses a free medium. This is because CSMA fails to detect a busy medium, thus there will be collisions at STA B. However, STA A cannot receive the collisions. In this context we say that STA A is hidden for STA C and vice versa.

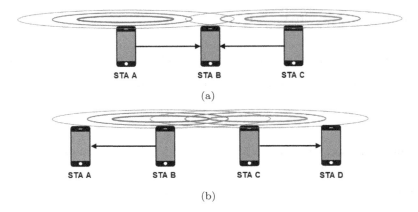

FIGURE 2.24

(a) Hidden Terminal Problem and (b) Exposed Terminal Problem

Figure 2.24(b) illustrates the **exposed terminal problem**, where we have four stations involved. In this scenario STA B transmits to STA A while STA C wants to transmit to STA D. When sensing the medium, STA C detects a medium in use and defers its transmission to STA D. However, STA A is outside the coverage area of STA C, therefore deferring the transmission is not necessary. Consequently, in this situation we say that STA C is exposed to STA B.

One solution to overcome these issues is **Multiple Access with Collision Avoidance (MACA)** [33] which uses two types of short signalling packets for collision avoidance, namely Request to Send (RTS) and Clear to Send (CTS). A station that wants to transmit it has to request the right to transmit to a receiver with a short RTS packet before it can being the data transmission. The receiver grants the right to send through a short CTS packet as soon as it is ready to receive. These packets contain the information regarding the sender and the receiver, as well as the expected duration of their communication. MACA avoids the hidden terminal problem illustrated in Figure 2.24(a) as follows: STA A and STA C want to transmit information to STA B, STA A transmits the RTS first and STA C waits after receiving the CTS from STA B. Collisions at STA B are avoided.

MACA also avoids the exposed terminal problem illustrated in Figure 2.24(b) as follows: STA B wants to transmit data to STA A while STA C wants to transmit data to another terminal STA D. However, by using MACA STA C does not have to wait as it cannot receive CTS from STA A. Thus, STA C communicates with STA D.

2.7 Practical Use-Case Scenario: Antennas Using Altair WinProp

One of the leading software in the area of wireless propagation and radio network planning is WinProp from Altair.[1] WinProp is in fact a collection of software consisting of: (1) **ProMan** which is the propagation manager and main simulator for wave propagation and radio network planning; (2) **Co-Man** which is the connectivity manager and simulator for sensor and mesh networks; (3) **WallMan** which is the wall manager and graphical editor for vector building databases; (4) **TuMan** which is the tunnel manager and graphical editor for tunnels and stadiums; (5) **AMan** which is the antenna manager and graphical editor for antenna patters; (6) **CompoMan** which is the editor for components used in wireless indoor network installations.

The aim of this practical use-case scenario is to use WinProp in order to investigate the radiation patterns of various antennas within an urban environment. To be able to achieve this, we are going to use first WallMan to generate a database map containing the geometry of a given urban area. AMan will be used to produce different antenna pattern files. Finally, the geometry from WallMan and the antenna pattern from AMan will both be used by the main radio propagation simulation tool ProMan.

2.7.1 Creation of Urban Database Using WallMan

The workflow in WallMan is as follows:

- convert/import the file from another tool (e.g., OpenStreetMap[2])

- make optional modifications: building shapes, towers, courtyards, vegetation, etc.

- save for use in ProMan

For the purpose of this practical use-case we have used OpenStreetMap to obtain a database file to be converted in WallMan. The urban area we have selected is located near Middlesex University in London, United Kingdom. However, it is possible to select any area you wish to investigate. The *.osm* file obtained from OpenStreetMap will be converted using WallMan (e.g., File > Convert Urban Database > Vector Database). Additional optional settings could be used like: check fill objects and check display building heights in status bar. Figure 2.25 illustrates the outcome of this step which can be saved as an *.odb* file and is ready to be used in ProMan.

---

[1] Altair: https://www.altair.com/
[2] OpenStreetMap: www.openstreetmap.org

FIGURE 2.25
WallMan Urban Database – Example

2.7.2 Produce 3D Antenna Pattern Using AMan

In order to generate propagation simulations we require antenna patterns to define signal sources. In this context, AMan can produce the antenna pattern files to be used in ProMan. The workflow for pattern import and conversion in AMan is as follows:

- import file

- convert two 2D patterns into one 3D pattern

- save for use in ProMan

AMan offers several techniques to produce a pattern file. In this practical use-case we are going to explore the use of two file formats, such as: *.msi* files and *.pln* files obtained from different antenna suppliers.

The first antenna to be considered is the **Huawei Antenna Model A90451702v01**, which is a single-band antenna, operating in the frequency range of 790–960 MHz, with a 3 dB horizontal beam width in the range of 63^0–67^0 and vertical beam width in the range of 8.9^0–9.8^0, a gain of 17 dBi, dimensions of 1936 x 260 x 135 mm and no electrical downtilt. The *.msi* file for this antenna was obtained from the MSI Antenna Pattern File Library available on the wireless planning website.[3] Figure 2.26 gives a visualization of this antenna including the horizontal and vertical patterns and the 3D representation generated by AMan.

To obtain the 3D representation the *.msi* file is imported in AMan through File > Convert MSI to 3D. By accepting the default conversion the 3D representation from Figure 2.26 is obtained.

Figure 2.27 illustrates the interpolated 3D patterns for the Huawei antenna, using four different interpolation algorithms available in AMan, such as: (1) Arithmetic Mean – makes use of the gain in the horizontal plane and in the vertical planes, the simplest algorithm considered to be the most inaccurate for this reason; (2) Bilinear Interpolation – where four gain values are weighted with their angle distances; (3) Weighted Bilinear Interpolation – similar to the arithmetic mean with additional weights considered for the

---

[3]https://www.wireless-planning.com/msi-antenna-pattern-file-library

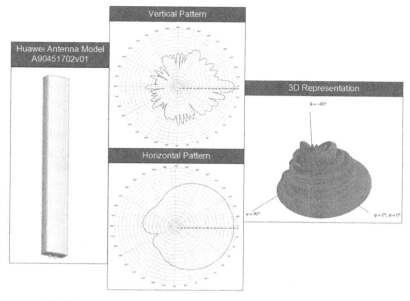

FIGURE 2.26
Visualization of Antenna Pattern

vertical angles which leads to more accurate results compared to the bilinear interpolation especially for antennas with the main radiation in the horizontal plane; (4) Horizontal Projection Interpolation – is mainly used for antennas with mechanical or electrical downtilt, using this method the gains in the horizontal and vertical planes are processed considering the gain of the horizontal pattern as a basis.

The second antenna to be considered is the **6 Element Yagi 890–960 MHz antenna – YB806-94**[4] belonging to the 800 MHz Yagi Antennas 540–1000 MHz YB806 Series. These antennas have a narrow bandwidth and high front to back ratios that are effective in reducing the interference with other systems.

The YB806-94 antenna operates in the frequency range of 890–960 MHz, a vertical beamwidth of 49^0, a horizontal beamwidth of 61^0 and a length of $0.6\,m$. Figure 2.28 gives a visualization of this antenna including the horizontal and vertical patterns and the 3D representation generated by AMan.

To obtain the 3D representation the $*.pln$ file is first opened in AMan and the 2D vertical and horizontal representations are saved. These are then used in AMan through File > Convert 2x2D to 3D. By accepting the default conversion the 3D representation from Figure 2.28 is obtained.

---

[4]https://www.rfi.com.au/YB806-94

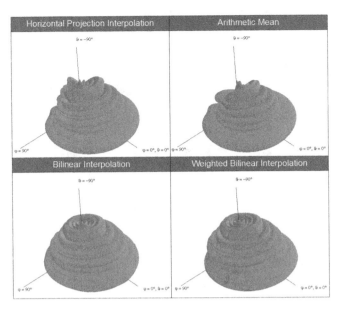

FIGURE 2.27
Various Interpolation 3D Patterns

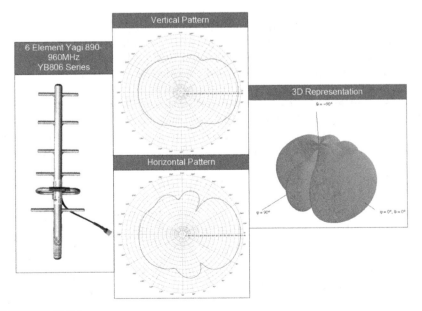

FIGURE 2.28
Visualization of Antenna Pattern

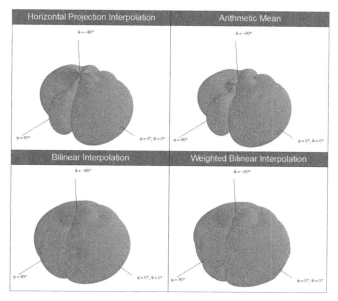

FIGURE 2.29
Various Interpolation 3D Patterns

Figure 2.29 illustrates the interpolated 3D patterns for the Yagi antenna, using four different interpolation algorithms available in AMan.

2.7.3 Urban Simulation Using ProMan

Several Antenna patterns were produced in AMan. The geometry representing a city was generated in WallMan. We will now use both in the propagation simulation tool ProMan to produce coverage plots and study the difference when using the two antennas with different interpolation 3D patterns deployed at the same location.

The overall workflow in ProMan is as follow:

- load geometry database that was prepared in WallMan.

- define source locations.

- specify antenna pattern that was prepared in AMan.

- set up simulation, including selection of a method.

- run simulation.

- inspect results.

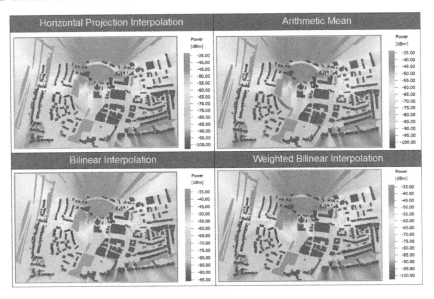

FIGURE 2.30
Radio Propagation for Huawei Antenna under Various Interpolation 3D Patterns

Figures 2.30 and 2.31 representing the predicted received power, show the impact of the antenna pattern on the computation of the wave propagation. The simulations were run considering the antenna positioned at a height of 10 m on top of a building of height 17.50 m. Thus, the antenna is positions at 27.50 m above the ground level. The frequency was set to 900 MHz and the *Dominant Path Model* was used as the propagation model. The predictions are computed at a height of 1.5 m above ground level.

The figures show similar results regardless of the interpolation being used, with the computations of the received power being within the same range. While comparing the radio wave propagation results of the two antennas, it can be noticed that there are visible differences concerning the main love and the side lobes. Consequently, the Huawei antenna has a wider beamwidth, providing a wider coverage at high power levels in the direction of the main lobe. On the other side, the Yagi antenna has stronger back lobes which enables it to provide better coverage at higher power levels towards the back, the opposite direction of the main lobe.

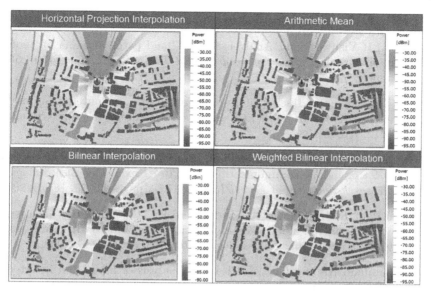

FIGURE 2.31
Radio Propagation for Yagi Antenna under Various Interpolation 3D Patterns

Bibliography

[1] William Stallings. *Wireless communications & networks: Pearson new international edition PDF eBook.* Pearson Higher Ed, 2013.

[2] Theodore S. Rappaport, Yunchou Xing, Ojas Kanhere, Shihao Ju, Arjuna Madanayake, Soumyajit Mandal, Ahmed Alkhateeb, and Georgios C. Trichopoulos. Wireless communications and applications above 100 ghz: Opportunities and challenges for 6g and beyond. *IEEE Access,* 7:78729–78757, 2019.

[3] Jochen H Schiller. *Mobile communications.* Pearson education, 2003.

[4] Cory Beard and William Stallings. *Wireless communication networks and systems.* Pearson, 2015.

[5] Behrouz A. Forouzan. *Data communications and networking.* McGraw-Hill, Inc., USA, 3 edition, 2003.

[6] Jerry D Gibson. *The communications handbook.* CRC press, 2018.

[7] Lal Chand Godara. *Handbook of antennas in wireless communications.* CRC press, 2018.

[8] Jerry D Gibson. *Mobile communications handbook*. CRC press, 2012.

[9] Babau R Vishvakarma. Analysis of dual-band patch antenna for mobile communications. *Microwave and Optical Technology Letters*, 47(6):558–564, 2005.

[10] RG Vaughan and J Bach Andersen. A multiport patch antenna for mobile communications. In *1984 14th European Microwave Conference*, pages 607–612. IEEE, 1984.

[11] Umair Rafique, Hisham Khalil, et al. Dual-band microstrip patch antenna array for 5g mobile communications. In *2017 Progress in Electromagnetics Research Symposium-Fall (PIERS-FALL)*, pages 55–59. IEEE, 2017.

[12] Binu Paul, S Mridula, CK Aanandan, and P Mohanan. A new microstrip patch antenna for mobile communications and bluetooth applications. *Microwave and Optical Technology Letters*, 33(4):285–286, 2002.

[13] Mohammod Ali and Gerard J Hayes. Analysis of integrated inverted-F antennas for bluetooth applications. In *2000 IEEE-APS Conference on Antennas and Propagation for Wireless Communications (Cat. No. 00EX380)*, pages 21–24. IEEE, 2000.

[14] W El Hajj, Christian Person, and Joe Wiart. A novel investigation of a broadband integrated inverted-f antenna design; application for wearable antenna. *IEEE Transactions on Antennas and Propagation*, 62(7):3843–3846, 2014.

[15] C Soras, M Karaboikis, G Tsachtsiris, and V Makios. Analysis and design of an inverted-f antenna printed on a pcmcia card for the 2.4 ghz ism band. *IEEE Antennas and Propagation Magazine*, 44(1):37–44, 2002.

[16] Y Belhadef and N Boukli Hacene. Pifas antennas design for mobile communications. In *International Workshop on Systems, Signal Processing and their Applications, WOSSPA*, pages 119–122. IEEE, 2011.

[17] Zhengwei Du, Ke Gong, Jeffrey S Fu, Baoxin Gao, and Zhenghe Feng. A compact planar inverted-f antenna with a pbg-type ground plane for mobile communications. *IEEE Transactions on Vehicular Technology*, 52(3):483–489, 2003.

[18] Pekka Salonen, L Sydanheimo, Mikko Keskilammi, and Markku Kivikoski. A small planar inverted-f antenna for wearable applications. In *Digest of papers. Third International Symposium on Wearable Computers*, pages 95–100. IEEE, 1999.

[19] Peter S Hall, Ee Lee, CTP Song, and R Waterhouse. Planar inverted-f antennas. In *Printed Antennas for Wireless Communications*, pages 209–218. Wiley, 2007.

[20] Sridhar Rajagopal, Shadi Abu-Surra, Zhouyue Pi, and Farooq Khan. Antenna array design for multi-gbps mmwave mobile broadband communication. In *2011 IEEE Global Telecommunications Conference-GLOBECOM 2011*, pages 1–6. IEEE, 2011.

[21] Xiaoming Chen, Muhammad Abdullah, Qinlong Li, Jianxing Li, Anxue Zhang, and Tommy Svensson. Characterizations of mutual coupling effects on switch-based phased array antennas for 5g millimeter-wave mobile communications. *IEEE Access*, 7:31376–31384, 2019.

[22] He-Sheng Lin and Yi-Cheng Lin. Millimeter-wave mimo antennas with polarization and pattern diversity for 5g mobile communications: The corner design. In *2017 IEEE International Symposium on Antennas and Propagation & USNC/URSI National Radio Science Meeting*, pages 2577–2578. IEEE, 2017.

[23] Jan Mietzner, Robert Schober, Lutz Lampe, Wolfgang H Gerstacker, and Peter A Hoeher. Multiple-antenna techniques for wireless communications-a comprehensive literature survey. *IEEE communications surveys & tutorials*, 11(2):87–105, 2009.

[24] M.A. Jensen and J.W. Wallace. A review of antennas and propagation for mimo wireless communications. *IEEE Transactions on Antennas and Propagation*, 52(11):2810–2824, 2004.

[25] Hyoungju Ji, Younsun Kim, Juho Lee, Eko Onggosanusi, Younghan Nam, Jianzhong Zhang, Byungju Lee, and Byonghyo Shim. Overview of full-dimension mimo in lte-advanced pro. *IEEE Communications Magazine*, 55(2):176–184, 2016.

[26] ChienHsiang Wu and Chin-Feng Lai. A survey on improving the wireless communication with adaptive antenna selection by intelligent method. *Computer Communications*, 181:374–403, 2022.

[27] Hrishikesh Venkataraman and Ramona Trestian. *5G Radio Access Networks: centralized RAN, cloud-RAN and virtualization of small cells.* CRC Press, 2017.

[28] Louis E Frenzel. *Principles of electronic communication systems.* McGraw-Hill (Boston), 2007.

[29] Wayne Tomasi. *Advanced electronic communications systems.* Prentice Hall PTR, 1997.

[30] Gordon L Stüber and Gordon L Steuber. *Principles of mobile communication*, volume 2. Springer, 1996.

[31] John D Day and Hubert Zimmermann. The osi reference model. *Proceedings of the IEEE*, 71(12):1334–1340, 1983.

[32] Norman Abramson. The aloha system: Another alternative for computer communications. In *Proceedings of the November 17–19, 1970, Fall Joint Computer Conference*, pages 281–285, 1970.

[33] Phil Karn et al. Maca-a new channel access method for packet radio. In *ARRL/CRRL Amateur Radio 9th Computer Networking Conference*, volume 140, pages 134–140. London, Canada, 1990.

Chapter 3

Radio Propagation

"A new, a vast, and a powerful language is developed for the future use of analysis, in which to wield its truths so that these may become of more speedy and accurate practical application for the purposes of mankind than the means hitherto in our possession have rendered possible."

Ada Lovelace
Algorithm Enchantress,
The World's First Computer Programmer

3.1 Introduction to Signal Propagation

Chapter 2 introduced the spectrum and frequencies used in wireless communications. We have seen that at low frequencies the signals can travel long distances and can penetrate objects and water easier. However, the achievable data rates are also low. On the other side, at high frequencies we could achieve higher data rates but at the cost of short distances as the signals cannot penetrate objects. Thus, **propagation** in wireless communication can be defined as the means the radio waves travel through free space or the medium, which in this case is represented by the air, from a source to a destination [1]. Depending on the way these waves travel their signal strength might be affected. In general, as the signal propagates through the medium, its signal strength will decrease with the distance. Consequently, several ranges could be defined around a sender as illustrated in Figure 3.1 [2], that is: (1) *transmission range* – representing the coverage area around a sender where transmission is possible, within this area the error rate is low enough for the sender and receivers to be able to communicate and for the receivers to act as transceivers. (2) *detection range* – represents the coverage area around the sender where the transmission is detected but the error rate is too high to be able to establish communication. (3) *interference range* represents the coverage area within which the receivers cannot detect signals but signals may disturb other signals, thus the sender may interfere with other transmissions.

DOI: 10.1201/9781003222095-3

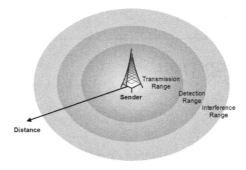

FIGURE 3.1
Transmission, Detection and Interference Range

Figure 3.1 represents just an ideal example of circular coverage area shape around the wireless transmitter tower. However, the real coverage would be represented by bizarrely-shaped polygons due to the dynamics of the environment and the existence of obstacles, such as buildings, mountains, hills, moving objects, etc [2].

Depending on their frequencies, the radio waves can travel from source to destination in three ways as indicated in Figure 3.2 [3; 4].

In the case of **ground wave propagation** (Figure 3.2(a)) the low frequency waves (e.g., < 2 MHz) can propagate over long distances and follow the curvature of the Earth. In the case of **sky wave propagation** (Figure 3.2(b)) the high frequency waves between 2 and 30 MHz, bounce between Earth's surface and the ionosphere to cover very large range. In the **Line-of-Sight (LoS)** (Figure 3.2(c)) propagation, the very high frequency waves (e.g., > 30 MHz) follow a straight line from the transmitter antenna to the receiver antenna without any obstacles in between.

However, as in wireless communications the radio signals propagate through the atmosphere, including air, rain, snow, fog, etc. there is a significant impact on the transmission especially over long distances [2]. Moreover, because in real life we might not always have LoS, the received power might be additionally influenced by various propagation effects, including: shadowing (e.g., through a wall), reflection at large obstacles, refraction depending on the medium density, scattering at small objects, diffraction at edges, etc. as illustrated in Figure 3.3.

- **Reflection** happens when the radio signal strikes against a very large dimensions obstacle as compared to the signal's wavelength. The obstacle would absorb some of the signal's power resulting in the reflected signal that is not as strong as the original. Consequently, the receiver might not be able to recognize the signals, because they became too weak if they have been reflected multiple times. However, in the environments where there is no LoS, reflections enable the transmissions of the signals.

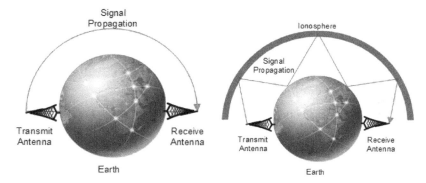

(a) Ground Wave Propagation (below (b) Sky Wave Propagation (2 to 30 MHz)
2 MHz)

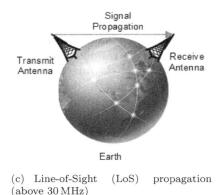

(c) Line-of-Sight (LoS) propagation
(above 30 MHz)

FIGURE 3.2
Propagation Methods

- **Diffraction** happens at the edge of an impenetrable obstacle that is large
  compared to the signal's wavelength. In this context, secondary waves
  propagate in different directions with the edge as the source. Due to this,
  the signal suffers a loss much greater than in the case of reflection. However, in the environments where there is no LoS, diffraction enables the
  transmissions of the signals.

- **Scattering** happens when the radio signal strikes against obstacles with
  dimensions smaller compared to the signal's wavelength, and the number
  of obstacles per unit volume is large (e.g., lamp posts, foliage, etc.). In this
  context, the signal is scattered in all directions resulting in reduced power
  levels.

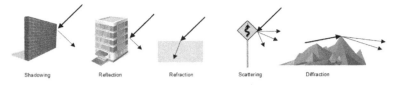

Shadowing   Reflection   Refraction   Scattering   Diffraction

FIGURE 3.3
Signal Propagation Effects

3.2 Multi-Path Propagation

In real life, the transmission medium between a transmitter and a receiver is more complex with many large or small obstacles including moving objects. In this context, due to reflection, scattering and diffraction the signals can travel to the receiver over various paths with different lengths and could arrive at the receiver at different times. Thus, the various copies of the signals that arrive at the receiver could be in phase and the direct signal could be enhanced, or they could be out of phase in which case the direct signal will be canceled out.

A simple example of a multi-path propagation scenario is presented in Figure 3.4 where a transmitter represented by a fixed antenna, is sending a narrow signal pulse at a given frequency to a mobile station. Each pulse will encode one or more bits of data. At the receiver side, it can be noticed that the first received pulse is the desired LoS signal that would travel on the shortest direct path. Because of the reflection and scattering one or more delayed copies of a pulse, referred to as secondary pulses or multi-path pulses may arrive at the same time as the primary pulse for a subsequent bit In this context, these secondary pulses act as a form of noise to the subsequent primary pulse causing **inter-symbol interference**. Moreover, as the mobile station moves, the location of various obstacles change, hence the number, magnitude, and timing of the secondary pulse will change as well. This effect caused by multi-path propagation is also called the **delay spread** that is a common effect in wireless communications.

3.3 Fresnel Zone

As we have seen in the previous section, there are several ways for a radio signal to travel between a transmitter and a receiver. The **Fresnel Zone** can be defined as a cylindrical ellipse that can be drawn around the direct LoS between a transmitter A and a receiver B, as illustrated in Figure 3.5 [5]. In case there are any obstacles located within the Fresnel zone, such as:

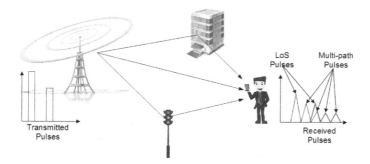

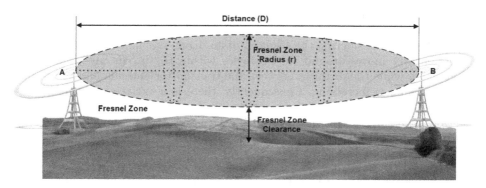

FIGURE 3.4
Multi-Path Propagation and Inter-Symbol Interference

FIGURE 3.5
Fresnel Zone

buildings, trees, hilltops, ground, etc. it could weaken the transmitted signal due to various propagation effects like reflection, diffraction, scattering, even if there is a direct LoS between the transmitter and the receiver.

It is possible to calculate the radius of the widest point of the Fresnel zone, located at half the distance between the transmitter and the receiver as seen in Figure 3.5 and given by the following equation [6]:

$$r = 8.657 \cdot \sqrt{D/f} \tag{3.1}$$

where r is the Fresnel zone radius in meters, D is the distance between the transmitter and the receiver in kilometers and f is the frequency in GHz. There can be an infinite number of Fresnel zones, however the primary Fresnel zone as defined above has the most impact on the performance of the wireless network. Consequently, in case there are no obstacles within the Fresnel zone between the transmitter and the receivers, the radio signals are transmitted without interference [7]. However, in case of obstacles or even the ground located within the Fresnel zone, the reflected signals could interfere with the direct LoS signal in a constructive or destructive way as previously explained. To improve the performance of the wireless system, it is recommended to keep the primary Fresnel zone at least 60% clear of any obstructions [8]. While for an optimal performance no more than 20% should be obstructed [9]. In case there is beyond 40% obstructions within the Fresnel zone, the signal strength become significant.

Consequently, for the best radio signal performance it is important to place the transmitter and the receiver antennas outdoors at a higher location in order to avoid the obstacles in the Fresnel zone.

As radio signals propagate through the environment they may penetrate through obstacles that appear in their path. However, due to the conductive nature of the various obstacles' materials, the radio signals might be more or less weakened or even completely blocked. The radio signals can be reflected by buildings or diffracted around sharp edges which weakness the strength of the signal.

However, considering the scenario of an urban mobile environment as the mobile user moves around the obstacles within the urban environment are changing while the distance between the transmitter antenna and the receiving antenna is increasing. Consequently, there is a variation in the average received power at the mobile user that is caused by shadowing and the variation in distance from the transmitter. This variation of the signal strength from its normal value is referred to as **fading** [4]. There are two types of fading identified as defined below:

- **long-term fading** that represents the changes in the mean value of the received signal strength, which is due to moving away from the transmitter and experiencing the path loss.

- **short-term fading** that represents the rapid variations in the signal strength, that are due to multi-path propagation and the doppler spread. These are caused by the speed of the mobile user, the transmission bandwidth and the environment.

3.4   Path Loss and Path Loss Models

As the radio signals travel from the transmitting antenna to the receiving antenna, their signal strength is attenuating over distance. The degree of attenuation is referred to as **path loss** and it is influenced by several factors including: the distance between the transmitting antenna and the receiving antenna, the ground and the obstacles on the path, the frequency components of the signals, etc. [10].

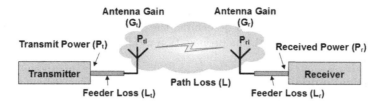

FIGURE 3.6
Wireless Communication System Elements

Figure 3.6 illustrates the elements of a simple wireless communication system [11] that contribute to the various losses and gains within the radio system making it hard to measure the path loss directly.

Consequently, the received power P_r can be defined as below [11]:

$$P_r = \frac{P_t \cdot G_t \cdot G_r}{L_t \cdot L \cdot L_R} \tag{3.2}$$

where P_t is the transmit power, G_t is the transmitter antenna gain, G_r is the receiver antenna gain, L_t is the transmitter feeder loss, L_r is the receiver feeder loss and L is the path loss. The antenna gain represents a measure for the directionality of an antenna and is defined as the power output, in a particular direction, compared to that produced in any direction by a perfect isotropic antenna.

The Effective Isotropic Radiated Power (EIRP) represents the maximum amount of power that could be radiated from an antenna, given its antenna gain, the transmitter power and the feeder loss. Thus, the effective isotropic transmit power P_{ti} is given by equation 3.3. While the effective isotropic received power P_{ri} is given by equation 3.4 [11].

$$P_{ti} = \frac{P_t \cdot G_t}{L_t} \tag{3.3}$$

$$P_{ri} = \frac{P_r \cdot L_r}{G_r} \tag{3.4}$$

Thus, the path loss L can now be defined as the ratio between the transmitted and received EIRP as given in the equation below:

$$L = \frac{P_{ti}}{P_{ri}} = \frac{P_t \cdot G_t \cdot G_r}{P_r \cdot L_t \cdot L_r} \tag{3.5}$$

In general, the path loss is expressed in decibels given by equation 3.6:

$$L_{dB} = 10 \cdot \log(\frac{P_{ti}}{P_{ri}}) \tag{3.6}$$

Various path loss propagation models have been defined to predict the received signal strength and to determine the coverage area of base stations and access points. In general, the propagation models can be classified in three categories [12; 13]:

- **empirical models** – are based on observations and measurements alone to describe the behaviour of a system. These models can be further split into time dispersive that provide information on the multi-path delay spread of the channel, and non-time dispersive that predict the mean path loss as a function of different parameters, including distance, antenna heights, etc.

- **deterministic models** – determine the received signal strength within a certain environment based on the laws governing electromagnetic wave propagation.

- **stochastic models** – make use of a series of random variables to model the environment and are the least accurate in their predictions.

3.5 Free Space Propagation Model

However, in real life it is unusual to have a LoS between a transmitter and a receiver due to the dynamic nature of the environment and the existence of obstacles. However, even if there are no obstacles between a receiver and a transmitter, the radio signals will attenuate over distance. This is known as **free space loss** and a simple free space equation has been defined by Friis [14] to denote the relationship between the transmission and the receive power as given in equation 3.7.

$$P_r = P_t \cdot G_t \cdot G_r \left(\frac{\lambda}{4 \cdot \pi \cdot d} \right)^2 \tag{3.7}$$

Where P_r is the received power, P_t is the transmit power, G_t is the gain of the transmitting antenna, G_r is the gain of the receiving antenna, d is the distance between the transmitting and the receiving antenna and λ is the wavelength given by $\lambda = f/c$.

The Friis free space equation is used for the prediction of the received signal strength when the transmitter and the receiver have a clear, unobstructed LoS path between them. The equation can also be written in terms of path loss in decibels where G_t and G_r are assumed to be 1, as follows [15][16]:

$$L_{dB} = 10 \log \left(\frac{P_r}{P_t} \right) = 10 \log \left(\frac{4 \pi d f}{c} \right)^2 = 20 \log \left(\frac{4 \pi d f}{c} \right)$$

$$L_{dB} = 32.44 + 20 \log(d) + 20 log(f) \tag{3.8}$$

where the frequency f is measured in MHz and the distance d in Km.

3.6   Two-Ray Ground Model

The two-ray ground model or the two-path model assumes that the radio signal travels from the transmitter to the receiver over two paths, one being the direct LoS and the second one is the path of the ground reflected signal [16]. In this context, the predicted received power can be calculated with the equation below:

$$P_r = P_t \cdot G_t \cdot G_r \cdot (\frac{h_t \cdot h_r}{d^2})^2 \tag{3.9}$$

where P_t is the transmit power, P_r is the received power, G_t and G_r are the transmitter and receiver antenna gains, respectively, d is the distance between the transmitter and the receivers and h_t and h_r are the transmitter and receiver heights, respectively.

It can be noted that as the distance increases the two-ray ground model shows a faster drop in the received power as compared to the free space model. Because the two-ray model is a multi-path propagation model, at short distances the model suffers of oscillations due to the constructive or destructive nature of the two rays. Consequently, when the distance is small the free space model is a better alternative [17].

3.7   Okumura Model

One of the first propagation models defined for urban areas is the Okumura model [16]. This model is applicable for operating frequencies in the range of 150 to 1920 MHz as well as distances of 1 km to 100 km. In contrast to the free space path loss, the Okumura model also takes into account the median attenuation relative to the free space (Amu) considering an effective antenna height at the base station (h_{te}) of 200 meters and the mobile stations antenna height (h_{re}) of 3 meters. Thus, the Okumura path loss model is given by equation 3.10 [10] below:

$$L_{50}(dB) = L_F + A_{mu}(f, d) - G(h_{te}) - G(h_{re}) - G_{AREA} \tag{3.10}$$

where L_{50} is the 50th percentile value of the propagation path loss, L_F is the free space propagation path loss, G is the gain factor and G_{AREA} is the gain due to the type of environment. It can be noted that the Okumura model consists of the free space path loss where the A_{mu} is added along with several correlation factors that account for the terrain type.

3.8 Okumura-Hata Model

The Okumura model was further improved by Hata [18], by fitting Okumura's curves with analytical expressions [16]. However, the formulation provided by Hata is limited to some set values of the input parameters. In this context, the median path loss in urban environments is given by the following equation [11]:

$$L_{50}(urban) = 69.55 + 26.16 \log f - 13.82 \log h_{te} - a(h_{re}) + (44.9 - 6.55 \log h_{te}) \log d \qquad (3.11)$$

Where f is the frequency with values in the range of 150 to 1500 MHz, h_{te} and h_{re} are the effective heights in meters of the base station and the mobile station antennas, respectively, d is the distance between the baste station and the mobile station, $a(h_{re})$ is the correction factor for the effective antenna height of the mobile station, which is a function of the size of the area of coverage. Consequently, for small to medium sized cities, $a(h_{re})$ in dB, is given by the following equation [16]:

$$a(h_{re}) = (1.1 \log f - 0.7)h_{re} - (1.56 \log f - 0.8) \qquad (3.12)$$

While for a large city, $a(h_{re})$ is given by [11]:

$$a(h_{re}) = \begin{cases} 8.29(\log(1.54h_{re}))^2 - 1.1, & f < 300\,MHz \\ 3.2(\log(11.75h_{re}))^2 - 4.97, & f \geq 300\,MHz \end{cases} \qquad (3.13)$$

The standard Hata formula can be modified to obtain the path loss in dB, in a suburban area, given by the equation below:

$$L_{50}(dB) = L_{50}(urban) - 2[\log(f/28)]^2 - 5.4 \qquad (3.14)$$

While the path loss in open rural areas is given by:

$$L_{50}(dB) = L_{50}(urban) - 4.78(\log f)^2 - 18.33 \log f - 40.98 \qquad (3.15)$$

The Okumura-Hata model is widely popular and is mainly used for large-cell mobile systems. However, it is not suitable for wireless communications systems with a circular coverage area of around 1 km in radius. Moreover, as the model is based on detailed measurement and analysis of the Tokyo area, it is somewhat specific to Japan's propagation environment. However, the Okumura-Hata model is one of the models that provides the best accuracy in a wide variety of situations.

3.9 COST 231 Walfisch Ikegami

The COST 231 Walfisch Ikegami (COST 231 W. I.) is an empirical model developed by the COST 231 project by combining the models from J. Walfisch and F. Ikegami [16]. The model operates in the frequency ranges from 800 to 2000 MHz and is primarily used in Europe for the GSM1900 systems [19; 20]. The model assumes that the building within a city are organized into nearly parallel rows similar to the Manhattan street grid where the streets are intersecting at right angles. While the precise heights and spaces of the buildings are ignored, the model takes into account the average rooftop level and the average distance separation between the rows of buildings, denoted as h_r and b, respectively [21; 22].

In this context, the median path loss in dB, is given by the sum of the free space loss L_f, rooftop-to-street diffraction and scatter loss L_{rts} and multi screen diffraction loss L_{msd} [13] as expressed below:

$$L_{50} = L_f + L_{rts} + L_{msd} \tag{3.16}$$

where the free space loss is defined by the equation below [19]:

$$L_f = 32.4 + 20 \log d + 20 \log f_c \tag{3.17}$$

The rooftop-to-street diffraction and scatter loss is given by:

$$L_{rts} = -16.9 - 10 \log w + 10 \log f_c + 20 \log(h_r - h_m) + L_{ori} \tag{3.18}$$

where w represents the width of the street between the buildings, h_m in meters is the heights of the mobile station, respectively, L_{ori} in dB represents the orientation loss given by [19]:

$$L_{ori} = \begin{cases} -9.646, & 0 \le \alpha \le 35^o \\ 2.5 + 0.075(\alpha - 35), & 35^o \le \alpha \le 55^o \\ 4 - 0.114(\alpha - 55), & 55^o \le \alpha \le 90^o \end{cases} \tag{3.19}$$

where α is the incident angle relative to the street. Finally, the multi screen loss is defined by [19]:

$$L_{msd} = L_{bsh} + k_a + k_d \log d + k_f \log f_c - 9 \log b \tag{3.20}$$

with

$$L_{bsh} = \begin{cases} -18 - 18(h_b - h_r), & h_b > h_b \\ 0, & otherwise \end{cases} \tag{3.21}$$

$$k_a = \begin{cases} 54, & h_b > h_b \\ 54 - 0.8(h_b - h_r), & r \ge 500; h_b \le h_r \\ 54 - 1.6r(h_b - h_r), & r < 500; h_b \le h_r \end{cases} \tag{3.22}$$

TABLE 3.1

Path Classes for Different Ray Classification

Path Class	Description
1	Direct path
2	Single reflection
3	Double reflection
4	Single diffraction
5	Triple reflection
6	One reflection + one diffraction
7	Double diffraction
8	Two reflections + one diffraction

$$k_d = \begin{cases} 18, & h_b > h_b \\ 18 - 15\frac{h_b - h_r}{h_r - h_m}, & otherwise \end{cases} \tag{3.23}$$

$$k_f = \begin{cases} 4 + 0.7(\frac{f_c}{925} - 1), & \text{for medium city and suburban area} \\ 4 + 1.5(\frac{f_c}{925} - 1), & \text{for metropolitan areas} \end{cases} \tag{3.24}$$

The COST 231 W. I. model is suitable for microcells and small macrocells [23]. As we have seen above, the model assumes constant building heights and uniform terrain, and when calculating the path loss it takes into account the free space loss, attenuation loss from the last roof edge to the mobile station as well as the multi screen loss that happens due to a series of rooftops perpendicular to the propagation path.

3.10 Intelligent Ray Tracing

The IRT is a three dimensional (3D) model that takes into account a maximum of three ray interactions, such as triple reflection and double diffraction in arbitrary combinations [24]. This combination is done according to eight specific classes as listed in Table 3.1 [25].

The following equation computes the total path loss of the IRT model taking into account the free space loss and the propagation path loss due to reflection and diffraction or penetration [22].

$$L_{total} = L_{fs} + G_{tx} + L_{interaction} \tag{3.25}$$

where G_{tx} is the directional gain of transmitting antenna in the direction of propagation path, $L_{interaction}$ is the loss in propagation path due to reflection, diffraction or penetration, while the free space loss L_{fs} is given by:

TABLE 3.2
Path Loss Exponent for Different Environments

Path Loss Exponent, n	Environment
2	Free space
2.7 to 3.5	Urban area cellular radio
3 to 5	Shadowed cellular radio
1.6 to 1.8	In building line-of-sight
4 to 6	Obstructed in building
2 to 3	Obstructed in factories

$$L_{fs} = 32.44 + 20 \cdot \log f + 10 \cdot n \cdot \log d \tag{3.26}$$

where f in MHz is the transmission frequency, d in km is the distance between the transmitter and the receiver, n is the path loss exponent. Typical values for n depending on the environment are given in Table 3.2

3.11 Dominant Path Model

Within the context of an urban scenario, in general there can be hundreds of possible propagation rays between a transmitter and a receiver. However, out of all these propagation rays, only two or three rays would account for more than 95% of the energy [22]. In this context, the Dominant Path Model (DPM) [26] has been defined to focus on the dominant rays only. DPM deals with urban, indoor as well as rural scenarios where the propagation loss for each path is determined deterministically based on the distance, path loss exponent and diffraction components. Consequently, the urban dominant path equation is given by [26]:

$$L = 20 \cdot \log(\frac{4\pi}{\lambda}) + 20 \cdot p \cdot \log d + \sum_{i=0}^{n} \alpha(\varphi, i) - \frac{1}{c} \sum_{k=0}^{c} w_k \tag{3.27}$$

where L is the path loss in dB, d is the path length, λ is the wavelength, p is the visibility factor for LoS and Non Line-of-Sight (NLoS) environments, $\alpha(\varphi, i)$ is the loss in dB due to an interaction and w_k is the wave guiding factor. More details on the algorithm and how the dominant path is decided can be found in [27].

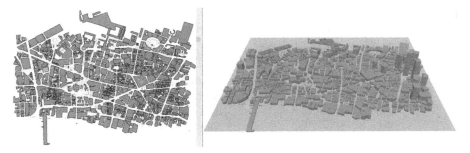

FIGURE 3.7
WallMan Urban Database – London Central Area Map

3.12 Practical Use-Case Scenario: Radio Propagation Using Altair WinProp

The aim of this practical use-case scenario is to use WinProp in order to investigate the effects of the propagation environment on the performance of a single-cell outdoor radio system and study the impact of various radio propagation models. To be able to achieve this, we are going to use first WallMan to generate a database map containing the geometry of a given urban area. Finally, the geometry from WallMan will be used by the main radio propagation simulation tool ProMan.

3.12.1 Creation of Urban Database Using WallMan

The workflow in WallMan is as follows:

- convert/import the file from another tool (e.g., OpenStreetMap[1])

- make optional modifications: building shapes, towers, courtyards, vegetation, etc.

- save for use in ProMan

For the purpose of this practical use-case we have used OpenStreetMap to obtain a database file to be converted in WallMan. The urban area we have selected is located in London central area, United Kingdom. However, it is possible to select any area you wish to investigate. The *.osm* file obtained from OpenStreetMap will be converted using WallMan (e.g., File > Convert Urban Database > Vector Database). Additional optional settings could be used like: check fill objects and check display building heights in status bar. Figure 3.7 illustrates the outcome of this step which can be saved as an *.odb*

---

[1]OpenStreetMap: www.openstreetmap.org

file. Normaly, this step would have been enough and the *.odb* database file would have been ready to be used in ProMan.

However, in this use case we are going to analyze the impact of three radio propagation models within an urban environment, namely COST 231 Walfisch-Ikegami (COST 231 W.I.), Intelligent Ray Tracing (IRT) and the Urban Dominant Path Model (DPM). Consequently, we need to pre-process the *.odb* file before using it in ProMan. For this a new project is created (File > New Project) that will use the *.odb* file we just created. We then use the Preprocessing > Edit Preprocessing Parameters option to select the mode for the pre-processing run for each propagation model. The output would be three different database files for each mode. Thus, the *.ocb* database file represents the Outdoor COST 231 W.I. Binary database, the *.oib* database file represents the Outdoor IRT Binary database and the *.opb* database file represents the Outdoor Urban DPM Binary database. These files are generated by selecting *Preprocessing > ComputeCurrentProject* in the menu and they contain building data as well as the visibility information along with the considered parameters for this pre-processing. These files are now ready to be used in ProMan with their corresponding propagation model.

3.12.2 Urban Simulation Using ProMan

The geometry representing the London central area was generated in WallMan with three different pre-processing database files, *.ocb*, *.oib* and *.opb* each corresponding to a propagation mode such as COST 231 W.I., IRT and the Urban DPM, respectively. We will now use all these database files in the propagation simulation tool ProMan to produce coverage plots and study the difference in path loss when using the three different radio propagation models within the same urban environment and the same single radio cell antenna characteristics deployed at the same location.

The overall workflow in ProMan is as follow:

- load an Urban Scenario

- load the preprocessed vector database that was prepared in WallMan.

- define source locations.

- specify antenna characteristics.

- set up simulation, including selection of a radio propagation model.

- run simulation.

- inspect results.

Figures 3.8 and 3.9 represent the predicted path loss and the field strength including an example of ray paths, for the three propagation models, COST

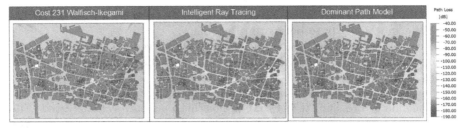

FIGURE 3.8
Path Loss Prediction for COST 231 W.I., IRT and Urban DPM for Antenna at 15 m

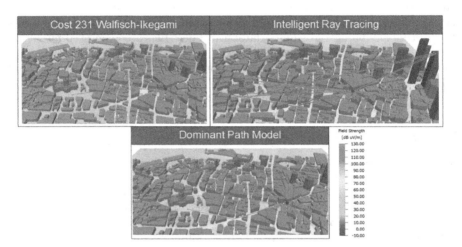

FIGURE 3.9
Field Strength Prediction and Ray Path Example for COST 231 W.I., IRT and Urban DPM for Antenna at 15 m

231 W.I., IRT and the Urban DPM, respectively. The simulations were run considering an omni-directional antenna, with a transmit power of 10 Watts and 947 MHz frequency, similar to the settings in [26]. In the first instance the antenna height considered was of 15 m above ground but below the rooftop level. The predictions are computed at a height of 1.5 m above ground level.

It can be noted that the building around the transmitting antenna are obstructing the direct rays. Consequently, the COST 231 W.I. model becomes too pessimistic, especially at distances further away from the transmitting antenna. This is mainly due to the fact that the model does not take into account the impact of diffraction and reflections that are the main factors for the wave-guiding effects on the streets. On the other side, the IRT model takes into account these effects and we can see better path loss prediction

FIGURE 3.10

Path Loss Prediction for COST 231 W.I., IRT and Urban DPM for Antenna at 30 m

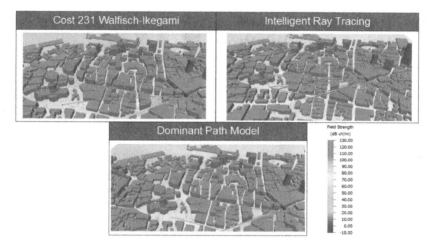

FIGURE 3.11

Field Strength Prediction and Ray Path Example for COST 231 W.I., IRT and Urban DPM for Antenna at 30 m

results. The multiple interactions considered by this model are also visible in Figure 3.9. However, as the distance from the transmitter increases the results become pessimistic, especially behind the buildings on the north-western part of the prediction area. This is because the fixed number of iterations limits the accuracy of the model especially if the rays found are not the ones carrying the main part of the energy [26]. These disadvantages are not visible in the case of the Urban DPM as this model shows comprehensive results across the prediction area. It can also be noted that in the case of both IRT and Urban DPM higher power values can be reached at the receiver for further distances from the transmitter as compared to the COST 231 W.I. model.

Figures 3.10 and 3.11 represent the predicted path loss and the field strength including an example of ray paths, for the three propagation models,

COST 231 W.I., IRT and the Urban DPM, respectively. However, the difference with the previous results is that this time the transmitter antenna is located at 30 m above ground and also above the rooftop level for most of the surrounding buildings. The rest of the settings were kept the same as above. It can be noted that for both transmitter antenna location at 15 m and 30 m a similar behaviour is observed between the three propagation models. However, the transmitter antenna located at 30 m provides an increase in coverage area for all three models.

3.13 Practical Use-Case Scenario: Rural/Suburban Study Using Altair WinProp

In the rural/suburban areas where there is a low density of buildings the radio propagation mainly depends on the topography and the land usage or clutter. ProMan contains a various propagation models for these types of scenarios [28]. The aim of this practical use-case scenario is to use WinProp in order to investigate the effects of different radio propagation models on the performance of a single-cell outdoor radio system within a rural/suburban environment. To be able to achieve this, we are going to use a detailed topographical database and the land usage or clutter database (water, roads, buildings, forest, etc.) of a hilly terrain area. We are going to use ProMan to study the impact of the topography and land usage on three propagation models of the electromagnetic waves, such as: Hata-Okumura model, Longley-Rice model and Rural Dominant Path Model (DPM).

The overall workflow in ProMan is as follow:

- load a Rural/Suburban Scenario and consider additional land usage (clutter, morpho)

- load the three databases for the hilly terrain: topography (pixel database), Land usage/Clutter (Pixel Database), Land usage/Clutter (Class definition).

- define source locations.

- specify antenna characteristics.

- set up simulation, including selection of a radio propagation model.

- run simulation.

- inspect results.

The environment considered is a hilly terrain and the map including the three databases are available from Altair Winprop. However, in general the

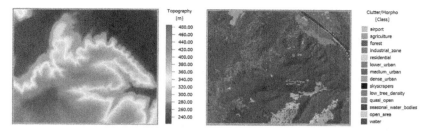

FIGURE 3.12

Topography and Clutter/Morpho (Land Usage) for the Rural/Suburban Scenario

topography and the clutter databases are usually supplied by external vendors. Figure 3.12 illustrates the topography (elevation) and clutter/morpho (land usage) describing the geometry of the hilly terrain of the rural/suburban use-case scenario. It can be noted from the land usage that the upper right area of the map is represented by the city and the industrial zone, the upper left is represented by medium and lower urban zone while the centre includes forested hills.

The same settings as the previous use-case were kept, with two omni-directional antennas, with a transmit power of 10 Watts and 947 MHz frequency, deployed in the area one on the highest hill and another one on the lower hill closer to the town zone. The antennas height considered was of $15\,m$ above ground. The predictions are computed at a height of $1.5\,m$ above ground level.

The received power predictions in dBm of the two transmit antennas are illustrated in Figure 3.13. In general, the standard radio propagation models for rural and suburban areas are based on empirical approaches and only the direct ray between the receiver and transmitter is computed which might not be the one carrying the main part of the energy [28]. Consequently, this often leads to too pessimistic results. Figure 3.13 shows a comparison between the received power predictions of the Hata-Okumura model, Rural DPM and Longley-Rice model.

The standard Hata-Okumura model is a simple approach that does not take into account the topography between the transmitter and receiver. However, in this simulation we have used the Hata-Okumura model with the *Individual Environment as defined in the morpho/clutter databases* option enabled. The results show that this option enables the model to take into account the terrain information. Even though the results are too pessimistic. On the other side, the Longley-Rice model has applications over rough terrain and includes diffraction, refraction and reflection interaction [24]. The results show too dominant effects of the considered and the model statistical approximation. This makes the shadows behind the hills to be too hard leading to too pessimistic results in this case as well. Instead, the results show that the Rural

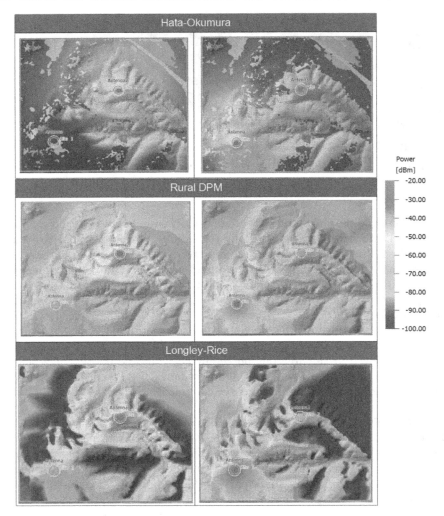

FIGURE 3.13
Received Power Prediction for Hata-Okumura, Rural DPM and Longley-Rice

DPM model which uses a full 3D approach for the path searching, obtains the more realistic and accurate results. Consequently the DPM model brings the following advantages [28]:

- the most important propagation path is computed by using a full 3D approach

- short computation times

- accuracy exceeds the accuracy of empirical models

Bibliography

[1] Donald E Kerr. *Propagation of short radio waves*, volume 24. IET, 1987.

[2] Jochen H Schiller. *Mobile communications*. Pearson education, 2003.

[3] Behrouz A. Forouzan. *Data communications and networking*. McGraw-Hill, Inc., USA, 3 edition, 2003.

[4] Cory Beard and William Stallings. *Wireless communication networks and systems*. Pearson, 2015.

[5] Martin PM Hall. Effects of the troposphere on radio communication. *Stevenage Herts England Peter Peregrinus Ltd IEE Electromagnetic Waves Series*, 8, 1980.

[6] Javad Ahmadi. The effects of fresnel zone in communication theory based on radio waves. *Bulletin de la Société Royale des Sciences de Liège*, 85:729–734, 2016.

[7] Hervé Sizun and Pierre de Fornel. *Radio wave propagation for telecommunication applications*. Springer, 2005.

[8] Rafhael Amorim, Preben Mogensen, Troels Sorensen, István Z Kovács, and Jeroen Wigard. Pathloss measurements and modeling for uavs connected to cellular networks. In *2017 IEEE 85th Vehicular Technology Conference (VTC Spring)*, pages 1–6. IEEE, 2017.

[9] DE Bassey, RC Okoro, and BE Okon. Issues associated with decimeter waves propagation at 0.6, 1.0 and 2.0 peak fresnel zone levels. *International Journal of Science and Research*, 5(2):159–163, 2016.

[10] Zahera Naseem, Iram Nausheen, and Zahwa Mirza. Propagation models for wireless communication system. *Signal*, 5(01), 2018.

[11] Michael S. Mollel and Michael Kisangiri. An overview of various propagation model for mobile communication. In *Proceedings of the 2nd Pan African International Conference on Science, Computing and Telecommunications (PACT 2014)*, pages 148–153, 2014.

[12] VS Abhayawardhana, IJ Wassell, D Crosby, MP Sellars, and MG Brown. Comparison of empirical propagation path loss models for fixed wireless access systems. In *2005 IEEE 61st Vehicular Technology Conference*, volume 1, pages 73–77. IEEE, 2005.

[13] Phillip A Thomas, Sylvester M Nabritt, and Madjid A Belkerdid. Propagation models used in wireless communications system design. In *Proceedings IEEE Southeastcon'98 'Engineering for a New Era'*, pages 230–233. IEEE, 1998.

[14] Harald T Friis. A note on a simple transmission formula. *Proceedings of the IRE*, 34(5):254–256, 1946.

[15] RA Valenzuela. Antennas and propagation for wireless communications. In *Proceedings of Vehicular Technology Conference-VTC*, volume 1, pages 1–5. IEEE, 1996.

[16] Hemant Kumar Sharma, Santosh Sahu, and Sanjeev Sharma. Enhanced cost231 wi propagation model in wireless network. *International Journal of Computer Applications*, 19(6):36–42, 2011.

[17] Theodore S Rappaport et al. *Wireless communications: principles and practice*, volume 2. prentice hall PTR New Jersey, 1996.

[18] Masaharu Hata. Empirical formula for propagation loss in land mobile radio services. *IEEE transactions on Vehicular Technology*, 29(3):317–325, 1980.

[19] Mardeni Bin Roslee and Kok Foong Kwan. Optimization of hata propagation prediction model in suburban area in malaysia. *Progress in Electromagnetics Research C*, 13:91–106, 2010.

[20] John David Parsons and Prof J David Parsons. *The mobile radio propagation channel*, volume 2. Wiley New York, 2000.

[21] Yihuai Yang, YongGang Xie, and Dongya Shen. An improved cost-wi model in metropolitan areas. 2009.

[22] SF Hamim and Mohd Faizal Jamlos. An overview of outdoor propagation prediction models. In *2014 IEEE 2nd International Symposium on Telecommunication Technologies (ISTT)*, pages 385–388. IEEE, 2014.

[23] Harsh Tataria, Katsuyuki Haneda, Andreas F Molisch, Mansoor Shafi, and Fredrik Tufvesson. Standardization of propagation models for terrestrial cellular systems: A historical perspective. *International Journal of Wireless Information Networks*, 28(1):20–44, 2021.

[24] Terhi Rautiainen, G Wolfle, and Reiner Hoppe. Verifying path loss and delay spread predictions of a 3d ray tracing propagation model in urban environment. In *Proceedings IEEE 56th Vehicular Technology Conference*, volume 4, pages 2470–2474. IEEE, 2002.

[25] R Hoppe. An introduction to the urban intelligent ray tracing (irt) prediction model. *AWE Communication GmbH*, pages 4–5, 2005.

[26] René Wahl, Gerd Wölfle, Philipp Wertz, Pascal Wildbolz, and Friedlich Landstorfer. Dominant path prediction model for urban scenarios. *14th IST Mobile and Wireless Communications Summit, Dresden (Germany)*, 2005.

[27] G Wolfle and Friedrich M Landstorfer. Dominant paths for the field strength prediction. In *VTC'98. 48th IEEE Vehicular Technology Conference. Pathway to Global Wireless Revolution (Cat. No. 98CH36151)*, volume 1, pages 552–556. IEEE, 1998.

[28] Altair. Altair winprop 2021.1 user guide. *Altair Engineering Inc.*, 2021.

Part II

Evolution of Mobile and Wireless Systems

Chapter 4

The Cellular Concept and Evolution

"My work on programming languages has always forced me to think in terms of what's the interface? Is this a nice, simple thing? How do you trade off the expressive power versus the ease of use and so on? And this is very useful in the systems domain, as well as in the programming language domain."

Dr. Barbara Liskov
Institute Professor at MIT,
Pioneer of Computer Programming

4.1 Cellular Systems Fundamentals

The cellular network is the underlying technology for mobile communications. It enables a Mobile Station (MS) to communicate with other fixed or mobile stations while stationary or on the move. Consequently, to make this possible the coverage area of a cellular network is divided into smaller regions called **cells**. A *cell* represents in fact the coverage area of a transmitting antenna that is controlled by a Base Station (BS). All MSs within the cell area are served by the BS and are able to communicate. Within an ideal radio environment the cell would have a circular shape around the BS as illustrated in Figure 4.1. However, in real live the shape of the cell depends on several factors, including the environment (e.g., mountains, hills, buildings, etc.), the weather conditions, the multi-path propagation and fading or even the system load [1]. For practical purposes, in theory the cell coverage area is approximated by a hexagon as seen in Figure 4.1. This is because a hexagon shape enables equidistant antennas and allows a larger region to be divided into non-overlapping hexagonal sub-regions of equal size, each representing a cell. This concept of cellular network consisting of cells it actually represents the implementation of Space Division Multiplexing (SDM) enabling high capacity in a limited spectrum system.

This is achieved by **frequency reuse**. The *frequency reuse* concept involves using the radio channels on the same radio frequencies to different areas that are physically separated from each other by considerable distance

DOI: 10.1201/9781003222095-4

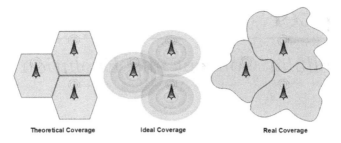

FIGURE 4.1
Cell Shapes

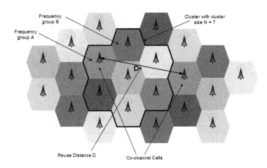

FIGURE 4.2
Frequency Reuse Concept, Cluster Size and Reuse Distance

to avoid interference. This brings the following benefits: (1) solves the problem of spectral congestion and user capacity by reusing the radio channels throughout the coverage region; (2) enables a fixed number of channels to serve an arbitrarily large number of users. Thus, offering high capacity in a limited spectrum allocation; (3) enables the use of low power transmitters instead of a single, high power transmitter (large cell) to cover larger area; (4) neighbouring BS are assigned different groups channels in order to minimize interference.

Figure 4.2 illustrates an example of the frequency reuse concept. The same colour cells use the same frequency group. A mobile service provider would have a limited set of frequencies available. Thus, to be able to offer communication to a large area, the frequencies have to be reused [2]. The frequency reuse pattern consisting of a number of N cells where each cell has a unique set of frequencies allocated, forms a **cluster**. In this context, N is the *frequency reuse factor* or the *cluster size*. This frequency pattern or the cluster is repeated to cover a specific service area of the cellular network. Consequently, the number of times the cluster is repeated is the same as the number off times the entire spectrum or bandwidth can be reused, which means increase in capacity. In the example provided in Figure 4.2, the frequency reuse factor or the cluster size is set to 7 ($N = 7$) and two cells using the same frequency group are separated by at least the *reuse distance D*. The radio channels that are reused by another cell physically located apart by at least D, are referred to as *co-channels*. Even though the BSs use some sort of power control to avoid interference, there still exists some degree of interference which is referred to as Co-Channel Interference (CCI). Different patterns or cluster sizes could be used. However, the most common one used in practice is with a cluster size of four and seven cells. As the cluster size increases, D becomes larger and the CCI decreases but the system capacity decreases as well. On the other side, as the cluster size decreases, D becomes smaller and the system capacity increases but the CCI increases. Regardless of the cluster size, there will always be six cells corresponding to the six sides of the hexagonal shape of a cell, using the co-channel at the reuse distance D. This is known as the first-tier co-channels. Moreover, the first-tier co-channels contribute to most of the CCI while the impact of the second-tier co-channels is negligible.

Capacity Enhancement – Worked Example (1)

Let us assume that we need to provide radio communication to a city. Our available bandwidth is 25 MHz and each user requires 30 KHz for a full-duplex communication channel. Compare the scenario where only one single high power antenna is deployed to the scenario where 20 low power antennas are deployed. How many simultaneous users can be served in both scenarios?

For *single antenna* scenario, the entire bandwidth of 25 MHz will be used by one high power antenna and the number of simultaneous users that could be served is the same as the number of available radio channels. Thus, this is given by taking the total available bandwidth of 25 MHz and divide it by the bandwidth required for one radio channels, which is 30 KHz, as calculated below:

$$\frac{25 \times 10^6}{30 \times 10^3} = 833 \ simultaneous \ users$$

For *20 low power antenna* scenario, we assume that the area is divided into five clusters with four cells per cluster ($4 \times 5 = 20$). Thus, the cluster size or the frequency reuse factor is equal to 4. This means that the total available bandwidth of 25 MHz will be used by one cluster consisting of 4 cells. Thus, we can calculate the amount of bandwidth allocated per cell as follows:

$$\frac{25 \times 10^6}{4} = 6.25 \ MHz$$

> While the number of simultaneous users per cell is given by:
>
> $$\frac{6.25 \times 10^6}{30 \times 10^3} \approx 208 \; simultaneous \; users$$
>
> Knowing that a cluster consists of 4 cells, we can calculate the total number of the simultaneous users per cluster as:
>
> $$4 \times 208 \approx 833 \; simultaneous \; users$$
>
> Meaning that the total system capacity is given by:
>
> $$5 \times 833 \approx 4165 \; simultaneous \; users$$
>
> representing *five* times increase in system capacity as compared to the scenario where only one high power antenna was used.

Based on the worked example provided above, a general formula for calculating the total capacity of a system in terms of the total number of simultaneous users is given below:

$$n = \frac{m \times BW_{CL}}{N \times BW_{ch}} \tag{4.1}$$

where n represents the total number of simultaneous users supported by the system (total capacity), m is the number of cells to cover the service area, BW_{CL} is the total available bandwidth per cluster, N is the frequency reuse factor (also known as the number of cells per cluster or cluster size), and BW_{ch} is the channel bandwidth.

Different techniques can be used to increase the system capacity, some of these techniques are listed below [3]:

(1) **Frequency Borrowing** – frequencies are taken (*borrowed*) from adjacent cells by congested cells.

(2) **Cell Splitting** – in areas with high traffic intensity, the cells can be split into smaller cells. Thus, cell splitting represents the process of subdividing a congested cell into smaller cells with their own BS in order to increase the system capacity and accommodate more traffic. Thus, more radio channels can be added without new spectrum usage. By deploying more smaller cells, this requires BSs with reduced transmit power and antenna height in order to keep the signal within the cell.

(3) **Cell Sectoring** – cells are divided into a number of wedge-shaped sectors, each with their own set of channels. This is achieved by replacing the omni-directional antenna by several directional antennas that would focus the radiation on a specific sector. In practice there are typically three or six sectors used per cell. Moreover, the use of directional antennas for cell sectoring reduces the CCI. For example, considering a system with three sectors per

cell, out of the six co-channels in the first-tier only two of them interfere with the central cell.

(4) Network Densification – represents the deployment of more cells of various sizes for frequency reuse. Consequently, according to the cell size, the cells can be classified into: *macro-cells* (e.g., 1 to 20 km radius) that cover a larger distance and can be used for the areas with low traffic density, for example rural areas where the radio coverage is served by a high power BS; *micro-cells* (e.g., 500 m to 2 km radius) where low power cellular BSs can be deployed to buildings, hills and lamp posts in areas with high traffic density whenever extra capacity is required; *pico-cells* (e.g., 50 to 500 m) are deployed to support even higher traffic density and cover small areas like street corner,offices, shopping malls, etc. and *femto-cells* (4 to 10 m) where low power cellular BS are deployed inside buildings to create the smallest coverage area cells.

(5) Interference Coordination – another way to increase the system capacity would be to employ a tighter control of interference so that frequencies can be reused closer to other BSs. Examples of interference control mechanism are: Inter-Cell Interference Coordination (ICIC) and Coordinated Multi-Point (CoMP).

Capacity Enhancement – Worked Example (2)

Let us assume a cellular system with a cell radius R of 1.6 km, a total of 32 cells, a total frequency bandwidth that supports 336 traffic channels, and a frequency reuse factor of $N = 7$. We need to evaluate the followings:

- What geographic area is covered by the system?

- How many channels are there per cell, and what is the total number of concurrent calls that can be handled by the system?

- Repeat for a cell radius of 0.8 km and 128 cells.

To calculate the total geographic area covered by the system, first we need to calculate the area of a cell. Knowing that the theoretical cell shape is a hexagon, we will calculate first the area of a hexagon as given by:

$$A = \frac{3\sqrt{3}}{2} \cdot R^2$$

for $R = 1.6\ km$, we get the area for a cell as $A = 6.65\ km^2$. As the system consists of 32 cells, the overall system area is $6.65 \times 32 = 213\ km^2$

In order to calculate how many channels are there per cell, we have the information on the total frequency bandwidth allocated for the cellular system that supports 336 traffic channels. These traffic channels are in fact the channels allocated per cluster that are repeated over the specific service area. Moreover, we know that the cluster size is $N = 7$, meaning that the cluster consists of 7 cells, we can calculate the number of channels per cell as follows:

$$\frac{336}{7} = 48 \ channels \ per \ cell$$

Knowing that the cellular system consists of 32 cells, we can calculate the total number of concurrent calls that can be handled by the system, which is the same as the total number of channels, calculated as:

$$48 \times 32 = 1536 \ concurrent \ calls \ or \ channels \ per \ system$$

We repeat the same steps for a cell radius reduced to half, $R = 0.8 \ km$ and an increase in the number of cells from 32 to 128 cells per system. Consequently, the area of the cell in this case would be $A = 1.66 \ km^2$ while the overall system area is $1.66 \times 128 = 213 \ km^2$. The cluster size and the total number of traffic channels per system remain the same, meaning that the number of channels per cell would be the same as in the previous case, give by $336/7 = 48$. However, the total system capacity would be given by $48 \times 128 = 6144 \ \ concurrent \ calls \ or \ channels \ per \ system$.

Thus, a reduction in cell radius by a factor of $1/2$ has increased the system capacity by a factor of 4.

4.2 Traffic Engineering in Cellular Systems – Problem Solving

When designing the telephone switches and the circuit-switching telephone networks, the traffic engineering concepts were developed and they also apply to cellular systems. In an ideal world, the number of available channels would equal the number of subscribers active at a certain time. However, in practice this it is not feasible for a cellular system to have enough capacity to handle any possible load at all times [3]. Considering a system that can handle a number of N simultaneous users given by the N number of channels, and the system had a total number of L subscribers. In this context, we can identify two system states:

- if $L \leq N$ then we deal with a *nonblocking system* where all the calls can be handled at any time

- if $L > N$ then we deal with a *blocking system* where a subscriber might be blocked because the capacity of the system is fully in use when attempting to make a call. The blockage might be represented by a busy tone.

Thus, when analyzing the performance of a blocking system, the following performance questions should be considered [3]:

- What is the probability that a call request is blocked?

- What capacity is needed to achieve a certain upper bound on probability blocking?

- What is the average delay until that call is put into service?

- What capacity is needed to achieve a certain average delay?

In practice, the cellular operators need to keep the average blockage rate below 2% [4]. One of the traffic engineering concepts is the **traffic intensity**, denoted by A which represents the load of the cellular system and is defined by [5]:

$$A = \lambda \times h \qquad (4.2)$$

where A is the traffic intensity measured in the dimensionless unit *Erlang*, λ is the mean rate of calls or connection requests per unit of time, and h is the mean holding time per successful call. Thus, the traffic intensity can be interpreted as the average number of calls arriving during the average holding period, for normalized λ.

Another traffic engineering concept is the **Grade of Service (GoS)** that represents a measure of the ability of the user to access a cellular system during the busiest hour. For example, a GoS of 0.01 can be translated into the fact that the probability that an attempted call is blocked during the busiest hour is equal to 0.01. In general, GoS values between 0.01 and 0.001 are considered good.

The blockage rate can be defined as a function of the number of subscribers, number of initiated calls and the length of the conversations or holding time. In this context, Erlang B equation is used to relate the call blocking probability P or GoS, to the average rate of the arriving calls and the average length of a call and is given by the formula below [6]:

$$P = \frac{\frac{A^N}{N!}}{\sum_{x=0}^{N} \frac{A^x}{x!}} \qquad (4.3)$$

where P is the call blocking probability (Grade of Service (GoS)), A is the traffic intensity in Erlangs, and N is the number of channels. Erlang B tables are readily available that could be used instead of the formula. An example of Erlang B table is given below.

TABLE 4.1
Erlang B Table [3]

	Capacity (Erlangs) for Grade of Service of				
Number of Channels, N	$P = 0.02$	$P = 0.01$	$P = 0.005$	$P = 0.002$	$P = 0.001$
1	0.02	0.01	0.005	0.002	0.001
4	1.09	0.87	0.7	0.53	0.43
5	1.66	1.36	1.13	0.9	0.76
10	5.08	4.46	3.96	3.43	3.09
20	13.19	12.03	11.10	10.07	9.41
24	16.64	15.27	14.21	13.01	12.24
40	31.0	29.0	27.3	25.7	24.5
70	59.13	56.1	53.7	51.0	49.2
100	87.97	84.1	80.9	77.4	75.2

Table 4.1 gives the details of three parameters, such as the Grade of Service (GoS) or call blocking probability P, the number of channels N and the traffic intensity A. Thus, given any two of the three parameters the third one can be computed using the table. Consequently, one could use the table to determine the capacity in terms of number of channels required to accommodate a given amount of traffic for a given GoS.

At a first glance, the Erlang B table demonstrates the two points below:

- *for a given GoS it is more efficient to have a larger capacity system than a smaller capacity one.* To demonstrate this point using the data in the Erlang B table, we can consider a single cell of capacity 10 channels that can handle 5.08 Erlangs at a GoS of 0.02. On the other side, for the same GoS of 0.02 and the same total capacity of 10 channels but this time offered by two cells with 5 channels capacity each a combined traffic intensity of 3.32 Erlangs can be handled.

- *larger capacity systems are more likely to be impacted by an increase in traffic.* To demonstrate this point using the data in the Erlang B table, we can consider a smaller capacity system with a single cell of capacity 5 channels that can handle 0.74 Erlangs at a GoS of 0.001. An increase in traffic of 49% will degrade the GoS to 0.005. However, when considering a larger capacity system with a single cell of capacity 40 channels, only an $\approx 11\%$ increase in traffic from 24.5 to 27.3 Erlangs results in a GoS degradation from 0.001 to 0.005.

4.2.1   Worked Example 01

A total of 210 MHz and 70 MHz of bandwidth are allocated to two cellular communication operators, Operator 1 and Operator 2, respectively for the deployment of their cellular systems within a city. The system of Operator 1 uses 30 kHz full-duplex channels and Operator 2 uses 25 kHz full-duplex channels to provide voice and control channels. The total allocated bandwidth for control channels for Operator 1 is 18 MHz, and for Operator 2 is 9 MHz respectively.

1. Assuming that for each system all cells have the same number of 800 voice channels for Operator 1 and 610 voice channels for Operator 2 in each cell, what is the frequency re-use factor for each Operator?

   Suggested Solution:

   Operator 1:

   – total bandwidth allocated for voice channels $= 210\ MHz\text{–}18\ MHz = 192\ MHz$

   – total number of voice channels per cluster $= 192\ MHz/30\ kHz = 6400$ voice channels/cluster

   – frequency reuse factor $= 6400/800 = 8$

   Operator 2:

   – total bandwidth allocated for voice channels $= 70\ MHz\text{–}9\ MHz = 61\ MHz$

   – total number of voice channels per cluster$= 61\ MHz/25\ kHz = 2440$ voice channels/cluster

   – frequency reuse factor$= 2440/610 = 4$

2. Suppose that in each system, the cluster of cells is duplicated 9 times. Find the total number of simultaneous communications that can be supported by each system.

   Suggested Solution:

   Operator 1:

   6400 voice channels/cluster $\times 9 = 57600$ simultaneous communications

   Operator 2:

   2440 voice channels/cluster $\times 9 = 21960$ simultaneous communications

3. What is the area covered in cells by each system?

   Suggested Solution:

   Operator 1:

   8 cells/cluster $\times 9 = 72$ cells/system

   Operator 2:

   4 cells/cluster $\times 9 = 36$ cells/system

4. Suppose the cell size is the same in both systems and a fixed area of 45 cells is covered by each system. Find the number of simultaneous communications that can be supported by each system.

Suggested Solution:

Operator 1:

800 voice channels/cell ×45 = 36000 voice channels/system (simultaneous communications)

Operator 2:

610 voice channels/cell ×45 = 27450 voice channels/system (simultaneous communications)

4.2.2 Worked Example 02

Consider a 3-cell system covering an area of $3100 \ km^2$. The traffic in the three cells is as follows:

Cell Number	1	2	3
Traffic (Erlangs)	30.8	66.7	48.6

Each user generates an average of 0.03 Erlangs of traffic per hour, with a mean holding time of 120s. The system is designed for a GoS of 0.02. Make use of the Erlang B Table in Table 4.1 defined above if needed.

1. Determine the number of subscribers in each cell and the number of calls per hour per subscriber.

Suggested Solution:

Number of subscribers = Traffic per cell/0.03.

Cell Number	1	2	3
Subscribers	1026	2223	1620

Number of calls per hour per subscriber, $\lambda = A/h = 0.03/(120/3600) = 0.9$

2. Determine the number of calls per hour in each cell and the number of channels required in each cell (hint: you will need to extrapolate using the Erlang B Table above).

Suggested Solution:

Number of calls per hour in each cell = number of subscribers ×0.9.

The Erlang B Table gives the value of A, use P = 0.02, and find N.

Cell Number	1	2	3
Calls per hour	923	2000	1458

Cell Number	1	2	3
Channels	40	78	59

3. Determine the total number of subscribers, the average number of subscribers per channel, and the subscriber density per km^2.

   Suggested Solution:

   Total number of subscribers = the sum of the values obtained from part (1) = $1026 + 2223 + 1620 = 4869$

   From (2), the total number of channels required = $40 + 78 + 59 = 177$

   Average number of subscribers per channel = total number of subscribers/total number of channels ≈ 28

   Subscriber density = total number of subscribers/3100 ≈ 92 subscribers per km^2

4. Determine the total traffic (total Erlangs) and the Erlangs per km^2

   Suggested Solution:

   Total traffic = the sum of the values from table in the problem statement = 146.1 Erlangs

   Erlangs per km^2 = $146.1/3100 \approx 0.047\ Erlangs/km^2$

4.2.3   Worked Example 03

Consider a 20 MHz one-way bandwidth available for a single operator, a channel bandwidth of 25 KHz and 12 control channels. The system has the following parameters: cell area = $2\ km^2$, total coverage area = $2000\ km^2$, cluster size = 12, average number of calls per user during the busy hour = 2.3, average holding time of a call = 80s, call blocking probability = 2%. Make use of the Erlang B Table in Table 4.1 above if needed to answer the following questions:

1. How many voice channels are there per cell?

   Suggested Solution:

   Total Number of channels per cluster = 20000 KHz/25 KHz = 800 channels/cluster

   Total Number of voice channels per cluster= 800 − 12 control channels = 788 voice channels/cluster

   Total Number of voice channels per cell = 788/12 (cluster size) = 65.66 voice channels/cell

2. Use the Erlang B Table and a simple straight-line interpolation to determine the total traffic carried per cell, in Erlangs/cell. Then convert that to Erlangs/km^2.

   Suggested Solution:

   The Erlang B Table gives the value of A, use P=0.02, and N $= 65.66$ to find out A as follows:

   $(65.66 - 40)/(70 - 40) = (A - 31)/(59.13 - 31)$;

   A $= 54$ Erlangs/cell;

   A $= 54/2\ km^2 = 27.45$ Erlangs/km^2

3. Calculate the number of calls/hour/cell, calls/hour/km^2, users/hour/cell and the number of users/hour/channel.

   Suggested Solution:

   Number of calls/hour/cell, $\lambda = 54/(80/3600) = 2430$ calls/hour/cell

   Number of calls/hour/$km^2 = 2430/2 = 1215$ calls/hour/km^2

   Number of users/hour/cell $= 2430/2.3$ (average number of calls per user during the busy hour) $= 1056$ users/hour/cell

   Number of users/hour/channel $= 1056/65 \approx 16$

4. A common definition of spectral efficiency with respect to modulation, or modulation efficiency, in Erlangs/MHz/km^2, is $\eta=$ Total traffic carried by the system/(Bandwidth $\times$ Total coverage area). Determine the modulation efficiency for this system.

   Suggested Solution:

   The total number of cells is 2000 km^2 (total system area)/2 km^2 (cell area) $= 1000$ cells/system

   $\eta = \frac{(54\ Erlangs/cell \times 1000\ cells)}{(20\ MHz \times 2000\ km^2)} = 1.25\ Erlangs/MHz/km^2$

4.2.4   Worked Example 04

Three different cellular systems share the following characteristics: frequency bands 824 to 844 MHz for mobile unit transmission and 869 to 889 MHz for base station transmission. A duplex circuit consist of one 25 kHz channel in each direction. The three systems have the cluster sizes of: 3, 12 and 19, respectively. Note that in each system, the cluster of cells (3, 12 and 19) is duplicated 4 times. Calculate the followings:

1. the number of simultaneous communications that can be supported by each system.

Suggested Solution:

We have the number of clusters $M = 4$;

bandwidth assigned to cluster $BW_{CL} = 40\ MHz$;

bandwidth required for each two-way channel $BW_{ch} = 50\ KHz$.

The total number of simultaneous calls or channels that can be supported by the system is given by $N = M \times BW_{CL}/BW_{ch} = 3200$ channels per system.

2. the number of simultaneous communications that can be supported by a single cell in each system.

Suggested Solution:

Total number of channels available per cluster is $N_{CL} = BW_{CL}/BW_{ch} = 800$. For a frequency reuse factor q, each cell can use $N_{cell} = N_{CL}/q$ channels per cell. Thus, for $q = 3$, $N_{cell} = 266$ channels/cell; for $q = 12$, $N_{cell} = 66$ channels/cell; $q = 19$, $N_{cell} = 42$ channels/cell.

3. the area covered in cells by each system.

Suggested Solution:

the area covered in cells is given by $N_{CL} \times q$. Thus, for system 1 $q = 3$, area is 12 cells/system; for system 2 $q = 12$, area is 48 cells/system, for system 3 $q = 19$, area is 76 cells/system.

4. Suppose the cell size is the same in all three systems and a fixed area of 40 cells is covered by each system. Find the number of simultaneous communications that can be supported by each system.

Suggested Solution:

the results obtained at part 2 $\times 40$. Thus, for system 1 we have 10640 channels/system, for system 2 we have 2640 channels/system and for system 3 we have 1680 channels/system.

4.3   Mobility Management and Handover

The mobility within the wireless and mobile environment can be classified according to the following aspects [7]: (1) *Terminal mobility* – the user's mobile device can change the Point of Attachment (PoA) wich represents the connection between the mobile device and the network, without interrupting the service; (2) *User mobility* – the user can access the service under the same identity, regardless of the PoA, or the mobile device (e.g., personalized SIM cards can be used for mapping the user to multiple devices); (3) *Service mobility* – the user can use a particular service regardless the mobile device used and the user location.

Efficient mobility management techniques are critical for the next generation networks as they need to provide some basic requirements [8]: support for all forms of mobility; mobility support for real-time and non-real-time applications; seamless user mobility support which enables the user to move within a heterogeneous wireless and mobile environment appertaining to different or the same service provider; the support for user's mobile device to change the PoA while moving without interruptions of the current session; the support for global roaming. Within a heterogeneous wireless systems, there are two types of roaming for mobile devices, such as [9]: (1) intra-system/intra-domain roaming which refers to the mobile device mobility between different cells of the same system (similar network interfaces and protocols) and (2) inter-system/inter-domain roaming which means that the mobile device can move between different technologies, protocols or service providers.

Some of the mobility protocols that can be used in order to enable the mobility at a global level are: Mobile IP version 4 (MIPv4) [10], Mobile IP version 6 (MIPv6) [11] and Proxy Mobile IPv6 (PMIPv6) [12] that is a network-based mobility management protocol which enables IP mobility without requiring the participation of the mobile device in the signaling process. The network handles the entire mobility process instead of the mobile device.

The mobility management consists of two main components, as illustrated in Figure 4.3: location management and handover management [13].

FIGURE 4.3
Mobility Management

The *location management* keeps track of the mobile device movement, in order to locate it in case of incoming calls, short messages or data. This service includes two main tasks [9]: (1) location registration or location update – the mobile device periodically informs the system about its current location, which in turn, maintains an updated location information database; (2) call delivery – in which case, when a communication request for the mobile device is initiated, the system has to determine the current location of the mobile device, based on the existing information in the databases.

The link between communication and mobility is enabled by the handover process [14]. A good definition of handover is given by ETSI and 3GPP [15] where the handover is defined as being the process by which the mobile device keeps its connection when changing the PoA (BS or Access Point (AP)). In terms of technologies, if both the source and target systems employ the same Radio Access Technology (RAT) and rely on the same specifications, then the handover process is referred to as **Horizontal Handover** [15]. If the target system employs a different RAT, then handover process is called **Vertical Handover** [16]. A comprehensive review on vertical handover solutions and their impact on the energy consumption is presented in [17]. The main objective of the handover process is to minimize the service disruption, which can be due to data loss and delay during the session transfer.

Figure 4.4(a) presents an example of the basic handover process: as a mobile user, on an ongoing call, gets further from BS A, its signal quality degrades due to mobility. As the mobile user nears the cell edge or border, it will leave the original coverage area of BS A and enter the coverage area of BS B. When the signal strength of BS B first exceeds that of the original cell, BS A, the handover process is triggered (location indicated by L_1 in Figure 4.4(a)), and the mobile terminal handovers to BS B. In this example, the handover decision is done based on the *Relative Signal Strength (RSS)* only. However, because the signal strength fluctuates due to multi-path propagation effects, this method can lead to a *ping-pong* effect in which the mobile terminal is repeatedly passed back and forth between two BSs. In Figure 4.4(a), the handover occurs at point L_1.

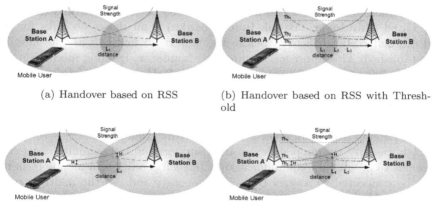

(a) Handover based on RSS

(b) Handover based on RSS with Threshold

(c) Handover based on RSS with Hysteresis

(d) [Handover based on RSS with Threshold and Hysteresis

FIGURE 4.4
Handover Management Decisions

Consequently, another option would be to base the handover decision on the *RSS with thresholds* as illustrated in Figure 4.4(b). In this case, the handover will occur only if the signal strength at the current BS A is less than a predefined threshold *and* the signal from a neighbouring BS B is stronger. Thus, the handover is avoided as long as the signal from the serving BS A is strong enough. However, for a high threshold value as indicated by Th_1 in Figure 4.4(b), this scheme performs the same as the RSS scheme. Whereas, if the threshold is set quite low as indicated by Th_3 the mobile may move far into the new cell. Consequently, finding the right threshold value is challenging.

Thus, another option would be to base the handover decision on the *RSS with hysteresis* as illustrated in Figure 4.4(c). In this case, the handover will occur only if the new BS B is sufficiently stronger, by a margin H than the current BS A. Thus, this scheme will trigger a handover when the RSS of the BS B reaches or exceeds H. Once the mobile terminal is handed over to BS B, it remains there until the RSS falls below $-H$, at which point it is handed back to BS A. In this way the *ping-pong* effect is prevented. However, the first handover may still be unnecessary if BS A still has sufficient signal strength.

A better approach would be to combine all these schemes and base the handover decision on the *RSS with thresholds and hysteresis* as illustrated in Figure 4.4(d). In this case, the handover will occur only if the current signal level drops below a threshold *and* the signal strength of the target BS B is stronger that the current BS A by a hysteresis margin H. Consequently, as indicated in Figure 4.4(d), the handover occur at location L_4 if the threshold is either Th_1 or Th_2 or at location L_3 if the threshold is at Th_3. Thus, this scheme avoids the *ping-pong* effect and also the execution of handover if the signal strength from the serving BS A is still strong enough.

The handover process can also be classified as illustrated in Figure 4.5: (1) **Hard Handover (Break-before-Make)** – all the radio links for the old original BS are removed before the new radio link(s) with the new BS is established. This approach requires fast handover signaling mechanisms in order to make the process transparent; (2) **Soft Handover (Make-before-Break)** – the new target link in the new BS is set-up first before the old original link from the old BS is torn down; (3) **Softer Handover** – it is a special case of soft handover. It may occur when a signal is replaced by a stronger signal from a different sector under the same BS.

Additionally the mobile device may support multiple simultaneous connections to be used for communications. In such conditions a mobile terminal may be connected to several BSs simultaneously and use the multiple radio links for wide bandwidth communications.

Handover Management consists of three major sub-services, as illustrated in Figure 4.6: Network Monitoring, Handover Decision, and Handover Execution [18].

Network Monitoring monitors the current network conditions as well as network availability. This service is responsible for gathering the data related to the network conditions, in order to trigger the handover execution when the service quality drops below the required QoS level. Relevant measurement criteria could include: received signal strength, distance, service quality in terms of error rates, traffic load, etc. Network Monitoring has to provide this gathered data, together with information related to the user preferences, current

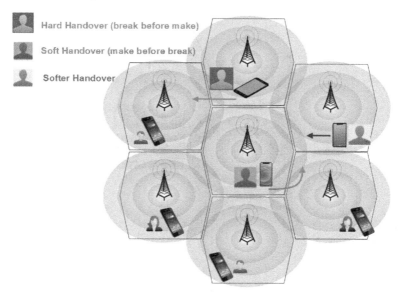

FIGURE 4.5
Types of Handover

running applications on the user's mobile device and their QoS requirements, to the Handover Decision module.

The *Handover Decision* handles the Network Selection process and is initiated either by an automatic trigger for a handover for an existing call or by a request for a new connection on the mobile device. The selection of the best network is decided based on the decision criteria provided by the device, the user inputs (if any), the application and the monitoring process. After the target network is selected the Handover Execution is triggered and the call is set up on the target candidate network. Traditionally, this network selection decision was made by network operators both for mobility and load balancing reasons, and was mainly based on a single parameter that is the received signal strength. However, there is now a wide range of network selection approaches proposed in the literature [19–22].

Every handover decision-making mechanism requires essential and relevant input information in order to choose the best value network as indicated in Figure 4.7. The decision criteria that may be used in the network selection process can be classified into four categories depending on their nature [13; 23]:

- *network metrics* – include information about the technical characteristics or performance of the access networks, such as: technology type, coverage, security, pricing scheme, monetary cost, available bandwidth, network load, latency, received signal strength, blocking probability, network connection time, etc.

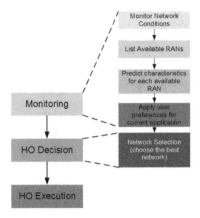

FIGURE 4.6
Handover Process

- *device-related metrics* – include information about the end-users' terminal device characteristics, like: supported interfaces, mobility support, capacity, capability, screen-size and resolution, location-based information, remaining battery power, etc.

- *application requirements* – include information about the requirements (minimum and maximum thresholds) needed in order to provide a certain service to the end-user: delay, jitter, packet loss, required throughput, Bit Error Rate, etc.

- *user Preferences* – include information related to the end-users' satisfaction: budget (willingness to pay), service quality expectations, energy conservation needs, etc.

An important aspect to consider is what information is readily available to the decision maker and how accurate and/or dynamic that information is. For example, because of the dynamics of the wireless environment the received signal strength or the available bandwidth can present major fluctuations for short periods; while coverage and pricing schemes are less dynamic as in they do not present changes on a daily basis; and technology type, security level and application requirements are more static parameters. Note that the parameters presented above do not represent an exhaustive list and are possible choices that can be used as input information for the decision mechanism. Some may use only a subset of the parameters, or may include additional parameters. Because the parameters present different ranges and units of measurement, they are normalized. The aim of the normalization process is to bring all the parameter into dimensionless units within $[0, 1]$ and make them comparable. The normalization process is done through the use of so called utility functions (normalization functions) [23]. The utility functions for the parameters may

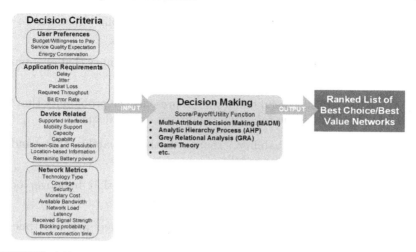

FIGURE 4.7
Decision-Making Process

vary. For example, some works consider normalized parameters based on the user and application requirements for the minimum and maximum value, while others consider normalization based on the ranges of values available from the different candidate networks. Other works consider using individual utility functions to model different parameters.

Due to the different possible strategies and the numerous parameters involved in the process, researchers have tried many different techniques in order to find the most suitable network selection solution, such as: game theory [23], Multi-Attribute Decision Making (MADM) [24], Analytic Hierarchy Process (AHP) [25], etc.

Handover Execution – after the target network is selected, the connection is set up on the target candidate network. In the case of an existing connection, the handover is executed, the original connection is torn down and the call data is re-routed to the new connection. If the first choice network is unavailable, then the next listed candidate is chosen as the target network. Connection setup (and teardown in the case of handover) will be handled by a mobility management protocol such as MobileIPv6. In order to provide good connectivity to the user the handover process has to be smooth, fast, seamless, and transparent. The main challenge in this process is to ensure the data does not get lost during the handover execution.

When assessing the performance of the handover process, several performance metrics should be considered:

- call blocking probability – as previously defined, it represents probability of a new call to be blocked due to high load on the BS traffic capacity. In this situation, the handover is performed to a neighboring BS due to the traffic capacity.

- call dropping probability – it represents the probability that due to handover a call is terminated.

- call completion probability – it represents the probability that an admitted call is not dropped before it terminates.

- probability of unsuccessful handover – represents the probability that handover is executed while reception conditions are inadequate.

- handover blocking probability – represents the probability that handover cannot be completed successfully.

- handover probability – represents the probability that handover occurs during current call.

- rate of handover – represents the number of handovers per unit time.

- interruption duration – represents the time length during handover in which the mobile unit is not connected to any PoA.

- handover delay – represents the duration it takes for the mobile user to move from the point where handover should occur to the point where it does occur.

4.4   Evolution from 1G to 5G and Beyond

This section introduces the evolution of mobile networks from 1G towards 5G and Beyond.

4.4.1   First Generation – 1G

The wireless cellular communications epoch started in the 1980s when the first mobile telephones (analogue phones) appeared. This first wireless cellular communication system was referred to as First Generation (1G). A good representation for this generation is the brick-sized analog phone intended to offer simple voice communication services to customers. Some examples of 1G systems deployed are [26]: Nordic Mobile Telephone 450 (NMT – 450) operated in the 450 MHz frequency range in Denmark, Sweden, Finland, and Norway; Total Access Communication System (TACS) at 900 MHz frequency range in United Kingdom; Advanced Mobile Phone Service (AMPS) operating within the 800 to 900 MHz frequency range in United States.

4.4.2 Second Generation – 2G

In the 1990s, the Second Generation (2G) of cellular systems emerged. Unlike 1G which used analog transmission for speech service, 2G used digital transmission. The second generation introduced, apart from the simple voice communication services, the low bit rate data services (e.g., SMS). Some of the 2G systems deployed are [26]: Global System for Mobile Communications (GSM) operating in the 900 MHz frequency range in Europe; Digital AMPS (D-AMPS) in United States; Code Division Multiple Access one (CDMAone) – based digital IS-95 in United States. GSM provided for interoperability of mobile devices between different operators leading to an easy and fast deployment of GSM all over the world. The GSM network is decentralized and consists of three separate subsystems [27]: Mobile Station (MS), Base Station Subsystem (BSS), and Network and Switching Subsystem (NSS), as illustrated in Figure 4.8.

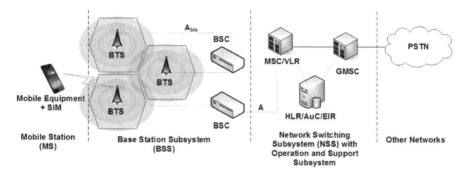

FIGURE 4.8
GSM Architecture

The MS is composed of the *Mobile Equipment* and the Subscriber Identity Module (SIM) card. The SIM card stores the subscriber's data, such as: identifiers, card type, serial number, list of subscribed services, Personal Identity Number (PIN), authentication key (K), PIN unblocking key (PUK), International Mobile Subscriber Identity (IMSI), compatible with locking of SIM in case of wrong PIN (for a three time trial). In the GSM network the SIM card is used to identify a mobile station in the network. The BSS is responsible for the radio network management and consists of two elements:Base Transceiver Station (BTS) and Base Station Controller (BSC). The BTS is usually placed in the centre of the cell and its transmitting power defines the coverage area of the cell. It is in charge of maintaining the air interface which is used for communication with the MS. The BSC is the main element of the BSS, and it is responsible to control the radio network. One BSC can control and manage the radio resources of a number of BTSs, and it is responsible with the mobility management of the MS (e.g., handover initiation).

The NSS consists of five elements: Mobile Switching Centre (MSC), Visitor Location Register (VLR), Home Location Register (HLR), Authentication Centre (AuC) and Equipment Identity Register (EIR). While the Operation and Support Subsystem (OSS) handles all functions related to networking operations and maintenance. The main role of theMSC is to control the calls in the mobile network and establish connections between BSCs and all other MSCs. The MSC can control several BSSs, and one BSS can cover a large geographical area consisting of many cells. A cell refers to the geographical area covered by one BTS. When the MSC acts as bridge between the mobile network and the fixed network (e.g., PSTN) it is referred to as Gateway MSC (GMSC). The VLR can be represented as an independent unit or it can be integrated within the MSC (MSC/VLR). VLR is a temporary database which holds information about the users (e.g., identity numbers, security information, subscribed services, originate HLR) as long as they are within the service area. HLR contains permanent information about the subscribers (e.g., identity numbers and subscribed services) along with the current location. EIR is used for security reasons, storing information about valid mobile equipment while AuC is responsible for the authentication and encryption parameters. Both entities, EIR and AuC, can be co-located with the HLR.

In terms of multiple access scheme, GSM uses a mix of FDMA and TDMA combined with frequency hopping (optional for the network operators). There are two frequencies bands of 25 MHz each and 45 MHz apart that have been reserved for GSM, such as: 890-915 MHz used for uplink communication indicating the MS to BS direction, and 935-960 MHz used for downlink communication indicating the BS to MS direction. Each of these bands of 25 MHz is sub divided into 124 single carrier channels of 200 kHz. Additionally, in each uplink/downlink bands guard bands of 200 KHz are used. Each 200 kHz channel carries 8 TDMA channels (8 time slots representing one TDMA Frame) with a duration of 577 μs per time slot (TDMA frame duration of 4.615 ms). The modulation used in GSM is GMSK which consists of a MSK procedure and the data is filtered through a Gaussian pulse shaping filter. In this case, the power spectrum of the signal becomes narrower, provides spectral efficiency, and good BER performance.

The mobility management in GSM consists of location management and handover management. The *location management* makes use of mechanisms to localize the users in case of incoming calls, SMS or data. There are two basic operations as part of the location management, such as: (1) *Location Update* which represents the operation initialized by the mobile station to inform the network about the user's location and (2) *Paging* – represents a broadcast message initialized by the network to locate the current cell of a user.

If the location management would be based on location updates only, that would mean that each time a user crosses the cell boundaries a location update is triggered. This leads to high signalling and database update overhead and high power consumption in the terminals. On the other side, if only paging is used, if a call arrives the mobile station is paged in all cells of the cellular system, leading to high signalling overhead and high delay in call delivery. Thus, to facilitate the location management process in an efficient way, the coverage area of a cellular system is partitioned into Location Area (LA)s that combine several cells. The size of a LA is determined by the cell radius, the mean mobile velocity, the cost of location updates overhead, the cost of paging overhead. The main goal when designing the location areas is to minimize the location management cost (location update + paging and processing).

A number of location update strategies exist, such as:

- periodic location updating – where the mobile station transmits location updates to the network periodically. However, the updates can be unnecessary if the mobile station does not move from a LA for a long time.

- location updating on LA crossing – where the BS periodically broadcasts the identity of its LA through a Location Area Identifier (LAI). As the mobile station permanently listens to the broadcast from the BS and stores the current LAI, it will check if the broadcasted LAI is different from the current LAI. In case they do not match, the mobile invokes the location update procedure. However, in the case of a highly mobile user, this method will generate a significant high number of location updates while for a low mobility user only few location updates are triggered.

- hybrid location updating – which combines the periodic and location updating on LA crossing mechanisms. In this case, the mobile station will generate the location update every time it detects an LA crossing. If there is no communication (related to a location update or a call) between the mobile station and the network for a fixed time period, the mobile station will generate a periodic location update.

Similarly, in terms of paging there are a few paging strategies available, such as:

- blank polling – this involves broadcasting paging messages on all BSs in LA simultaneously. Even though this method introduces a short delay, it has a high paging load.

- sequential polling – this method involves trying different sets of BSs sequentially, where the BS with the highest likelihood is selected first (e.g., the BS from which the last location update was issued). Even though this method introduces a less paging load, there is a higher delay observed.

Consequently, GSM is using a *hybrid location updating* process which involves: (1) periodic execution – executed by the mobile station if a timer, set by the operator, has expired and (2) execution on LA crossing.

There are two types of processes that are defined in GSM, such as: (1) *Mobile Terminated Call* – where a calling station located outside the GSM network calls a mobile station located within the GSM network and (2) *Mobile Originated Call* where a call is initiated by a calling station within the GSM network towards the outside.

The step involved in the *Mobile Terminated Call* are listed below [1]:

1. the user of the calling station dials the phone number of the GSM subscriber

2. the PSTN of the calling station notices that the number belongs to a mobile user in the GSM network and forwards the call setup to GMSC.

3. GMSC identifies the HLR for the subscriber (from the phone number) and signals the call setup to the HLR

4. HLR checks whether the number exists and whether the user has subscribed for the requested services, and requests a Mobile Station Roaming Number (MSRN) from the current VLR.

5. at the receipt of the MSRN, the HLR determines the MSC responsible for the MS and forwards this information to the GMSC.

6. GMSC forwards call setup request to the MSC.

7. MSC requests the current status of the MS from the VLR

8. if MS status is available, the MSC initiates paging in all the cells it is responsible for

9. BTS transmits the paging signal to MS, if the MS answers, the VLR performs the security check

10. VLR signals to MSC to setup the connection to the MS

The step involved in the *Mobile Originated Call* are listed below [1]:

1. MS transmits a request for a new connection

2. BSS forwards the request to the MSC

3. MSC checks with the VLR of the user is allowed to set up a call with the requested service

4. MSC checks the availability of resources through the GSM network and into the PSTN

5. if all the resources are available the MSC sets up the connection between the MS and the fixed network

The handover in GSM is an inter-frequency hard handover, as the adjacent cells always use different frequency ranges. There are different types of handover defined based on the involvement relation of the BSC and MSC entities. Consequently:

- Intra-BTS handover – where the handover occurs between different radio channels, within the same cell. Same BSC and MSC are maintained.

- Intra-BSC handover – where the handover performed between two BTS (cells) served by the same BSC. Same BSC and MSC are maintained.

- Intra-MSC handover – where the handover is performed between two BTS served by different BSC. MSC controls the handover. Same MSC is maintained.

- Inter-MSC handover – where the handover is executed between two BTS served by two different MSCs.

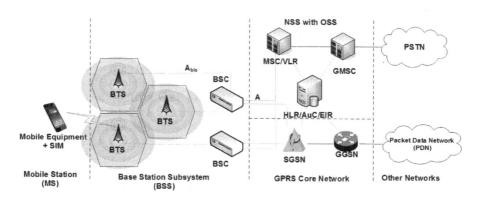

FIGURE 4.9
GPRS Architecture

4.4.3   2.5 Generation – 2.5G

Based on the GSM system, new and more advanced technologies were developed [28]: (1) High Speed Circuit Switched Data (HSCSD) offers higher data transmission rates, the theoretical maximum data rate being 57.6 Kbps; (2) General Packet Radio Service (GPRS) which introduces a higher theoretical data rate of 160 Kbps; (3) Enhanced Data for Global Evolution (EDGE) brings further increases in data rates being able to handle multimedia services (e.g., video phone, video conference, etc) at a theoretical rate of 384 Kbps. GPRS adds support for packet switched data by integrating into the GPRS Core Network two main and new entities, as illustrated in Figure 4.9: the Serving GPRS Support Node (SGSN) and the Gateway GPRS Support Node (GGSN). The SGSN entity is the most important element of the GPRS network being equivalent to the MSC of the GSM network. Thus, it provides the same functions for the GPRS network as what the MSC provides for the GSM network. It can control one or more BSCs and it is responsible for delivery of packets to and from the MSs located within its service area. It is also in charge for the mobility management, security, authentication and charging functions. The GGSN entity represents the gateway that connects the GPRS network to the external networks. It also maintains routing information related to a MS, so that it can route packets to the SGSN servicing the MS. As illustrated in Figure 4.9 the GPRS network uses and works in parallel with the GSM network. Consequently, there are three modes of operation defined for mobile stations, such as: (1) Class A – MS that can be connected to both GPRS and GSM services at the same time; (2) Class B – MS can be attached to both GPRS and GSM services but they can only use one service at a time. For example, a Class B MS can make or receive a voice call, or send and or receive a SMS message while the GPRS service is suspended but it is re-established once the voice call or SMS session is completed. (3) Class C – MS can be attached to either GPRS or GSM services but user needs to switch manually between the two different types.

The location management is optimized in GPRS by the introduction of Routing Area (RA)s that are actually GSM Location Area (LA)s subdivided into several RAs. Similarly to GSM where the MS needs to be located for incoming calls, within GPRS network the MS needs to be located for the delivery of data packets. Consequently, the Routing Area Identifier (RAI) is used instead of the LAI. RAs are introduced in order to avoid paging the MS for every downlink packet within all the cells of the MSs' LA, which could lead to high overhead and high delay in the network. Thus, in GPRS routing areas and a state model for adaptive location management are used. The RAs are significantly smaller than LAs and depending on the GPRS state model location updates and paging are related to RA or cells. The MS can be in one of the three states: IDLE, READY and STANDBY. When the MS is in the IDLE state, it is not reachable in GPRS mode and the location management is done according to the GSM mode. When the MS is in the READY state, the location update is performed by the MS at the entering of a new cell. When the MS is in the STANDBY state, the location update is performed by the MS whenever entering a new RA.

The deployment of EDGE networks does not require major changes in the core network, exept for the installation of EDGE-compatible transceiver units. However, the BSS needs to be upgraded to support EDGE. EDGE is using a combination of GMSK and/or 8PSK as the modulation scheme over the air interface to provide higher data rates. In GPRS the symbol rate is 270.833 kbps with each symbol representing 1 bit. Whereas, EDGE uses 8PSK coding scheme with each group of 3 bits representing one symbol. Thus, the symbol rate is still the same as 270.833 kbps, but with each symbol now representing 3 bits, the data rate is effectively tripled to 812.5 kbps.

4.4.4   Third Generation – 3G

The further growth of the data traffic led to the deployment of the Third Generation (3G) of mobile networks which comes to offer higher data rates of up to 2 Mbps. The 3G mobile networks were standardized as the International Mobile Telecommunications 2000 (IMT-2000). The new standard is designed for

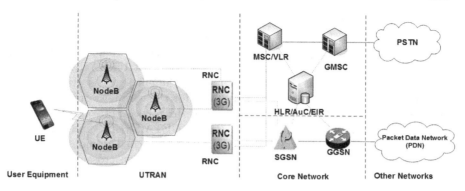

FIGURE 4.10
UMTS Architecture

internet/data services and low bit rates multimedia services. Examples of 3G systems are [29]: Wideband CDMA (WCDMA) developed from GSM and led by the Third Generation Partnership Project (3GPP), a joint project of the standardization bodies from different European countries; CDMA2000 – developed from 2G CDMA standard IS-95 in North America and Asia Pacific; Time Division – Synchronous CDMA (TD-SCDMA) in China. The most important 3G cellular system is Universal Mobile Telecommunications Service (UMTS) which uses WCDMA for the air interface. The principle behind WCDMA is that all users are simultaneously transmitting in the same frequency bands at the same time with each user interfering with each other. Thus, the cluster size is 1, as the adjacent cells use the same frequencies. However, the cells, User Equipment (UE), and physical channels are separated by codes. These codes consist of: (1) *channelization codes* that are used for separation of physical channels in the uplink and separation of UE in the downlink, called Orthogonal Variable Spreading Factor (OVSF), and (2) *scrambling codes* for separation of UE in the uplink and cells/sectors in the downlink.

UMTS keeps the concepts and solutions of the GSM network but a new infrastructure is required. The UMTS architecture is illustrated in Figure 4.10 and is composed of three main domains [27]: UE, UMTS Terrestrial Radio Access Network (UTRAN), and the Core Network (CN). The UE is the equivalent of the MS in GSM, with added support for UMTS. The Core Network is based on the GSM/GPRS network upgraded in order to support UMTS operation and services. The UTRAN provides the air interface for the UE, and is the equivalent of BSS in GSM, consisting of two main entities: NodeB and Radio Network Controller (RNC). NodeB is the equivalent of BTS whereas RNC is the equivalent of BSC. A RNC can control one or more NodeBs and performs radio resource management, mobility management, data encryption/decryption, etc.

The experiences from GPRS indicate that the location management is exclusively controlled in the core network (e.g., by SGSNs) while the procedures (paging and location/cell updates) must pass the interface between access and core network which lead to high load and large delays. Consequently, a new approach is introduced in UMTS which tracks the subscribers on the basis of RA in the core network and on the basis of UTRAN Registration Area (URA) and cells in the access network. The efficiency is also improved by the introduction of two state models, one for the core network and one for the access network. Consequently, there are three core network states, such as: (1) Packet Mobility Management (PMM) DETACHED – indicating that there is no location management in the packet switched domain; (2) PMM CONNECTED – the location updates are performed on crossing of RA and this state corresponds to the CELL CONNECTED or URA CONNECTED state in the access network; (3) PMM IDLE – indicating that the location updates are performed on crossing of RA.

Within the access network, there are three states defined as well, such as: (1) IDLE – there is no location management in the access network; (2) – CELL CONNECTED – the location updates are performed on crossing of cells and this state corresponds to the PMM CONNECTED state in the core network; (3) URA CONNECTED – the location updates are performed on crossing of URAs and this state corresponds to the PMM CONNECTED state in the core network.

UMTS make use of a power control mechanism such that all users in a cell experience the same Signal to Interference Ratio (SIR) at the NodeB receiver and avoids the near-far effect. The NodeB instructs the UEs closer in, to reduce their transmitted power, and those further away to increase theirs such that all UEs will be received at approximately the same strength. The *open Loop*, representing the initial power setting, is used during the initial access before the fully establishment of the communication between the UE and NodeB and it measures the received signal strength and estimates the transmitter power required. The *closed loop* representing the fast power control, is used when the UE has accessed the system and is in communication with the NodeB. In this case, the signal strength measurement is taken at each time slot and a power control bit is sent to request the power to be stepped up or down.

UMTS provides peak data rates of up to 384 kbps for uplink and downlink. After the NobeB functionality was upgraded to High-speed Downlink Packet Access/High-speed Uplink Packet Access (HSDPA/HSUPA) [30] the data rates could reach up to 14.4 Mbps for downlink and 5.76 Mbps for uplink.

The evolution towards the 3.5 generation of the cellular system leads to the deployment of the Evolved HSPA (HSPA+). The new HSPA+ offers improved data rates of up to 21 Mbps for downlink and up to 11 Mbps for uplink. This is because the NodeBs may be directly connected to the GGSN over a standard Gigabit Ethernet and the use of MIMO is introduced.

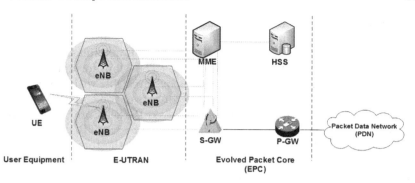

FIGURE 4.11
LTE Architecture

4.4.5 Fourth Generation – 4G

As mobile devices became a must on the market and also with the emerging of new multimedia applications that require more stringent QoS requirements, the deployment of Fourth Generation (4G) mobile networks attracted more and more interests. Thus, the next step in the GSM/UMTS evolution line is the 3GPP LTE [31]. LTE uses the OFDMA and Single-Carrier FDMA (SC-FDMA) transmission schemes for downlink and uplink, respectively, instead of WCDMA. The advantages of using new access techniques means that the receivers are simplified while the interference between adjacent cells is reduced. By using OFDMA, it enables the UEs to transmit at the same time over the sub-channels allocated to them. Thus, for OFDMA the users are allocated in time and frequency domains so that each user is only using a part of the available bandwidth making this scheme very flexible. Additionally, LTE introduces the flexibility of utilizing different bandwidth blocks, such as: 1.4 MHz, 3 MHz, 5 MHz, 10 MHz, 15 MHz, and 20 MHz. LTE comprises a maximum of 2048 different sub-carriers having a spacing of 15 KHz and the sub-carriers are split into Resource Block (RB)s with each RB consisting of 12 sub-carriers regardless of the overall LTE signal bandwidth. A RB covers one slot in the time frame and different LTE signal bandwidths will have different number of RBs (e.g., 1.4 MHz – 6 RBs, 3 MHz – 15 RBs, 5 MHz – 25 RBs, 10 MHz – 50 RBs, 15 MHz – 75 RBs, 20 MHz – 100 RBs). The disadvantage of using OFDMA is that it has a high Peak to Average Power Ratio (PAPR), that is not a problem for the BS but it is unacceptable for UE side due to resource constraints. However, in order to reduce the PAPR and increase the efficiency of the power amplifier and save battery life at the MS side, LTE uses a different access mode for the uplink, that is SC-FDMA. In the case of OFDMA each sub-carrier is modulated by a different data symbol that lasts for relatively long duration. On the other side, in the case of SC-FDMA, which in fact is a multi-carrier scheme but all the sub-carriers in the uplink are modulated by the same data, with a much shorter symbol duration.

The low PAPR of SC-FDMA means that it has a near constant power level when operating leading to increase in battery lifetime at the UE side.

Moreover, MIMO technology is used for the LTE antennas, providing data rates of up to 100 Mbps for downlink and up to 50 Mbps for uplink. The LTE system requires a new infrastructure that is incompatible with GSM or UMTS network, as illustrated in Figure 4.11. LTE consists of three main domains [27]: UE, Evolved-UTRAN (E-UTRAN) and Evolved Packet Core (EPC). The UE requires to be upgraded for LTE compatibility. The E-UTRAN consists only of evolved NodeB (eNB)s and is responsible for all the functionalities of the radio interface. EPC consists of three main entities: the Mobility Management Entity (MME) which handles the mobility, security and UE identity provided by Home Subscriber Server (HSS) that stores all user-related and subscriber-related information; the Serving Gateway (S-GW) which receives and sends packets between the eNB and the core network, and Packet Data Network Gateway (P-GW) with connects the EPC with external networks. Thus, LTE has a simplified architecture, being a flat IP-based network it does not have support for circuit-switched voice. Instead network operators can use Voice over LTE (VoLTE). A new soft frequency reuse scheme was also introduced where the inner part of the cell uses all the subbands with less power, while the outer part of the cell uses the pre-served subbands with higher power.

One of the most important functions of the E-UTRAN is handled by the Radio Resource Management (RRM) entity that covers the uplink and downlink functionalities like radio admission control, mobility, packet scheduling and dynamic radio resources allocation. The scheduling process is done by the eNB in order to satisfy the QoS requirements for a certain traffic category. LTE uses bearers (e.g., defined as the tunnels used to connect the UE to Packet Data Network (PDN)) for QoS control instead of circuits and each bearer is given a QoS Class Identifier (QCI). As the end-to-end service is not completely controlled by the LTE network, this is split between the Evolved Packet System (EPS) bearer connecting the UE to the P-GW and the external bearer. The EPS bearer maps to specific QoS parameters, such as data rate, delay, packet error rate, etc. In this regard, the radio bearers can be divided into two main classes: (1) Guaranteed Bit Rate (GBR) bearers – dedicates radio resources in order to guarantee a minimum bit rate and possibly higher bit rates if there are system resources available. Types of services that belong to this class are: voice, interactive video or real-time gaming; (2) Non-Guaranteed Bit Rate (Non-GBR) bearers – does not guarantee a minimum bit rate and the performance is more dependent on the number of UEs served by eNB as well as the system load. Types of services that belong to this class are: e-mail, file transfer, web browsing, etc.

The LTE eNB packet scheduler is responsible for achieving the QoS requirements for each radio bearer which are identified based on the QCI as per Table 4.2. Each QCI is represented by resource type, priority, packet delay budget in [ms] and packet loss rate.

TABLE 4.2

Standardized QCI Characteristics

QCI	Resource Type	Priority	Packet Delay Budget [ms]	Packet Loss Rate	Service Examples
1	GBR	2	100	10^{-2}	Conversational Voice
2	GBR	4	150	10^{-3}	Conversational Video (live streaming)
3	GBR	3	50	10^{-3}	Real Time Gaming
4	GBR	5	300	10^{-6}	Non-Conversational Video (buffered streaming)
5	Non-GBR	1	100	10^{-6}	IMS Signalling
6	Non-GBR	6	300	10^{-6}	Video (buffered streaming), TCP-based (e.g., www, e-mail, chat, ftp, file sharing, progressive video, etc.)
7	Non-GBR	7	100	10^{-3}	Voice, Video (live streaming), Interactive Gaming
8	Non-GBR	8	300	10^{-6}	Video (buffered streaming), TCP-based (e.g., www, e-mail, chat, ftp, file sharing, progressive video, etc.)
9	Non-GBR	9	300	10^{-6}	Video (buffered streaming), TCP-based (e.g., www, e-mail, chat, ftp, file sharing, progressive video, etc.)

Thus, through the packet scheduling process the radio resources are assigned to each user in order to enable QoS provisioning for the requested services in an efficient way. Two main concepts are involved in the LTE packet scheduling process, such as [32]: (1) the *scheduling rule* which is applied at every Transmission Time Interval (TTI) for the entire duration of the transmission and (2) the *scheduling procedure* which performs user selection and resource blocks allocation. The objective of the packet scheduling process is to maintain an acceptable trade-off between four main system parameters, such as [33]: capacity (e.g., system throughput, spectral efficiency, cell coverage), QoS, stability (e.g., robustness) and user fairness. This means that one could try to improve the performance of one or more of these system parameters without a significant impact on the rest of the objectives. However, this is hard to achieve by applying one single scheduling rule across the entire duration of the transmission especially under highly dynamic networks like the mobile networks. Consequently, many solutions in the literature have formulated the packet scheduling problem as a multi-objective optimization problem [33; 34] while the use of reinforcement learning will select the most suitable scheduling rule to be applied at each TTI from a set of scheduling rules [35–38], in order to find the best trade-off between the system parameters.

In terms of handover management, LTE makes use of packet forwarding to prevent data loss during the handover process. To this end, the source eNB decides to trigger the handover by sending a *handover command* message when the signal of the neighbouring eNB is stronger than the current signal at the UE. The source eNB would forward packets to the target eNB that buffers and transfers the undelivered data after the handover process is completed.

Parallel to the GSM/UMTS evolution line, the 3GPP2 Ultra Mobile Broadband (UMB) is the next successor of CDMA2000 [39]. UMB is based on OFDMA, MIMO and SDMA advanced antenna techniques, providing data rates up to 280 Mbps for downlink and over 75 Mbps for uplink transmission (e.g., this can be obtained by using 4x4 MIMO configuration).

The next advancement known as LTE Advanced (LTE-A), comes to bring advanced QoS capabilities, improved latency reduction, broader bandwidth, wider coverage area, smooth handover, etc. The LTE-A is expected to reach data rates of 1Gbps for stationary users, and up to 100 Mbps for mobile users. Some of the key improvements are:

- support for larger bandwidth of 100 MHz and asymmetric transmission mode

- carrier aggregation – LTE-A signal can consist of an aggregation of multiple LTE carriers referred to as Component Carrier (CC) that can be of different bandwidth, such as 1.4, 3, 5, 10, 15 or 20 MHz. LTE-A can combine up to 5 CCs up to a total of 100 MHz. These CCs can be aggregated using three approaches: (1) Intra-band contiguous where the carriers are adjacent to each other; (2) Intra-band non-contiguous where multiple CCs belonging to the same band are used in a non-contiguous manner and (3) Inter-band non-contiguous where the aggregation is done across multiple bands.

- Inter-Cell Interference Coordination (ICIC) alleviates the data rate degradation of the UEs located at the cell edges due to inter-cell interference by using the Fractional Frequency Reuse (FFR) which separates the frequency bands and allocates the band efficiently to prevent signal interference from adjacent eNBs.

- Dynamic Sub-carrier Assignment (DSA) is an enhanced resource allocation scheme that dynamically allocate sub-carriers based on the channel state conditions and avoiding frequency selective fading.

- Coordinated Multi-Point (CoMP) aims to improve the coverage of high data rate cell-edge throughput, and system throughout by coordinating multiple eNB to communicate with an UE. n this regard, several CoMP technologies could be employed, such as: (1) *Coordinated Scheduling* – which allocates different sub-carriers to the UEs located at the edge of the cell to avoid inter-cell interference; (2) *Coordinated Beamforming* – which allocates different beam patterns to the UEs located at the edge of the cell to avoid interference and improve reception performance; (3) *Joint Transmission* – improves the reception performance by enabling the UE to receive data concurrently from several eNBs; (4) *Dynamic Point Selection* – selects the transmission point (eNB) with better channel quality to improve the reception performance.

- heterogeneous networks involving small cells such as femtocells, picocells for network densification in order to expand the network capacity. However, this will introduce new challenges, such as: increased number of handovers, frequency reuse, QoS provisioning and security.

- relay nodes typically deployed at the edge of the cell, aiming at extending the coverage area and increase capacity and throughput. A donor eNB will be used to communicate and connect the relay node to the rest of the network. The relay nodes can be setup inband using the same frequencies as the donor eNB or outband using different frequencies. From the UE perspective the relay node is a fully fledged eNB, while the donor eNB will abstract the relay node from the rest of the network.

- MIMO enhancements to support higher dimensional MIMO. Consequently, in LTE-A antenna configurations of 8x8 MIMO for downlink and 4x4 MIMO for uplink are proposed.

A comparison summary of the 3G to 4G technologies is provided in Table 4.3.

TABLE 4.3
Requirements UMTS vs. HSPA vs. HSPA+ vs. LTE vs. LTE-A

	UMTS	**HSPA**	**HSPA+**	**LTE**	**LTE-A**
Max downlink speed	384 Kbps	14 Mbps	28 Mbps	300 Mbps	3 Gbps
Max uplink speed	128 Kbps	5.7 Mbps	11 Mbps	75 Mbps	1.5 Gbps
Latency RTT	150 ms	100 ms	50 ms	$< 5\ ms$	$< 1\ ms$
Access Methodology	CDMA	CDMA	CDMA	OFDMA/ SC-FDMA	OFDMA/ SC-FDMA

4.4.6 Fifth Generation – 5G

The current digital transformation has seen a tremendous increase in the demand for multimedia-rich applications including AR, VR and Mixed Reality (MR) that cannot be handled by the current underlying infrastructure. Consequently, this led to the deployment of the Fifth Generation (5G) mobile networks, referred to as 5G New Radio (NR). In order to ease a smooth deployment progress from 4G to 5G two functionalities of the NR were introduced, such as [40]: **standalone NR** – where all the control and data plane functions are provided in the NR and **non-standalone NR** – where the control plane functions of the LTE/LTE-A networks are utilized as an anchor for NR. A simplified architecture of 5G is illustrated in Figure 4.12, where the eLTE eNB is the evolution of LTE eNB that enables the connectivity to EPC as well as the 5G Next Generation Core (NGC) and the next generation NB (gNB) is the new 5G base station that supports the NR and enables the connectivity to NGC. Thus, the new RAN can support both the eLTE eNB of the E-UTRAN and the gNB of the NR to interface with the NGC.

Figure 4.12 also illustrates two non-standalone deployment options such as: *non-standalone NR* – where the gNB requires the anchor LTE eNB for control plane connectivity to the EPC and *non-standalone E-UTRAN* where the eLTE eNB requires the anchor gNB for control plane connectivity to the NGC.

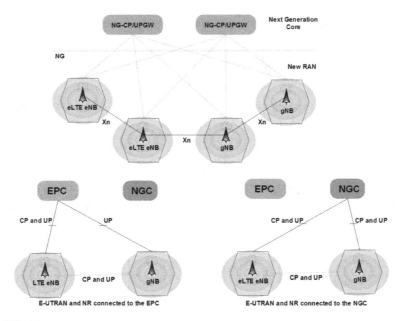

FIGURE 4.12
5G Architecture

Consequently, several deployment scenarios of the 5G NR could be envisioned as in Figure 4.13 [40]. In the first scenario presented in Figure 4.13(a) the 5G NR is deployed as an overlay over the LTE/LTE-A of the same coverage or as 5G NR small cells collocated with LTE/LTE-A eNB or non-collocated with LTE/LTE-A eNB. The second scenario depicted in Figure 4.13(b) considers the LTE/LTE-A eNB as a master node with the anchor LTE/LTE-A eNB offering control and user plane functionalities while the NR gNB acting as a booster. In this case, the data flow will aggregate via the EPC across both the LTE/LTE-A eNB and the gNB. Figure 4.13(c) presents a third possible deployment scenario where the NR gNB is the master node. In this situation a standalone NR gNB could offer control and user plane functionalities via the NGC. While it is also possible to have a collocated eLTE eNB to provide additional booster functionalities for dual connectivity. In this case, the data flow will aggregate via the NGC across both the eLTE eNB and the gNB. The fourth example illustrated in Figure 4.13(d) considers the eLTE eNB as a master node. Similarly, a standalone eLTE eNB could offer control and user plane functionalities via the NGC. While it is also possible to have a collocated NR gNB to provide additional booster functionalities for dual connectivity. In this case, the data flow will aggregate via the NGC across both eLTE eNB and the gNB. Finally, Figure 4.13(e) depicts a fifth deployment scenario illustrating the handover process between the 4G and 5G Radio Access Technology (RAT). In the case where the LTE/LTE-A eNB is connected to the EPC and the NR gNB is connected to the NGC the coordination between EPC and NGC is required to enable the handover. However, if an eLTE eNB is used instead and an NR gNB are both connected to the NGC, then the handover between the eLTE eNB and the gNB could be fully managed by the NGC.

According to Cisco [41] the primary contributors to the global mobile traffic growth are the different mix of wireless devices, including smartphones, M2M, tablets, Personal Computers, etc. It is estimated that by 2023 there will be 4.4 billion M2M connections represented by GPS in cars, asset tracking systems in shipping and manufacturing sectors, as well as medical applications for patient records and health status. The advances in IoT, VR, AR, ML, AI paves the way for future applications, such as: self driving vehicle diagnostics, UHD VR, cloud gaming, UHD streaming, etc. However, these multimedia-rich applications will require strict QoS requirements that need to be supported on a heterogeneity of hardware platforms associated with the content access devices and dynamic network conditions which might hamper their potential [42]. Among the major objectives of 5G, one is to enable the QoS provisioning among three different types of service classes, such as [40]:

- enhanced Mobile Broadband (eMBB) – supports high capacity and high mobility radio access for multimedia-rich applications like VR, 3D videos, work and play in the cloud, etc.

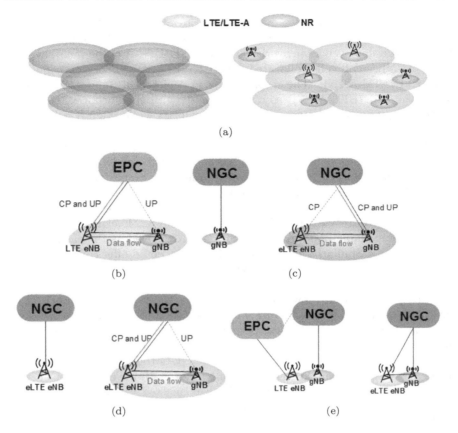

FIGURE 4.13
Deployment Scenarios of NR

- Ultra-Reliable and Low Latency Communications (URLLC) – provides urgent and reliable data exchange for mission critical applications like e-health, self driving cars, industrial automation, etc.

- massive Machine Type Communications (mMTC) – infrequent, massive and small packet transmissions for mMTC applications to enable connectivity for millions of devices per km^2 (e.g., smart cities)

This is achieved in 5G through a perfect storm of multiple technology breakthroughs [43] including *Network Slicing, Software-Defined Networking (SDN), Network Function Virtualization (NFV), beamforming, massive MIMO, mmWave, MEC, AI-based resource allocations*, etc. as illustrated in Figure 4.14.

Moreover, in order to improve the system capacity in 5G, some advanced physical layer techniques can be integrated, such as [40]:

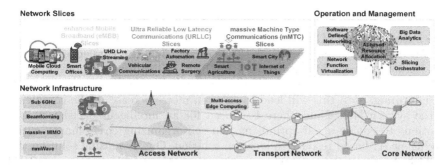

FIGURE 4.14
5G Key Enabling Technologies

- higher order modulation and coding scheme such as 256-quadrature amplitude modulation (256QAM);

- beamforming and mMIMO [44] where a very high number of antennas can be used to multiplex messages for several UEs on each time-frequency resource allocation. The radiated energy from the antennas could be focused towards the intended direction while also minimizing intra and inter-cell interference;

- ability to operate in any frequency band, including low band frequencies and sub-6GHz as well as the mmWave [45] that uses the spectrum between 24 GHz and 100 GHz which has a very short wavelength and is designed for short-range services and specific applications;

- new access schemes like: *Non Orthogonal Multiple Access (NOMA)* [46] – where the users are distinguished by the power levels so that users with pool channel conditions get higher power. At the receiver side, the users with higher power levels will decode their signal by treating others as noise while the users with lower power levels will subtract the higher powered signals before decoding their own signal. Another option is the use of *Filter Bank Multi Carrier (FBMC)* [47] that makes use of a filter to remove the sub-carrier overflow and eliminate the side lobes. In this context different users could have different sub-bands with different parameters;

- interference management with Successive Interference Cancellation (SIC) can be used at the receiver side to decode two or more overlapping signals which enables the use of multiple concurrent transmissions [48]. Moreover, by combining the use of SIC and NOMA the overall throughput could be improved by up to 30% when compared to OFDMA;

4.4.6.1   AI-Based Resource Allocation

The initial phase of 5G it is still using OFDMA in downlink and uplink, with SC-FDMA as optional for uplink. This means that the challenges faced by the Radio Resource Management (RRM) are still present. However, as 5G sees the integration of various hybrid emerging technologies (e.g., mMIMO, beamforming, etc.) this will lead to an even more increase in the complexity of the system [49]. Thus, there is a need for an intelligent decision-making solutions to enable a self-optimizing and self-organizing environment with a intelligent AI-based resource allocation solution. One promising solution that could boost network performance is the integration of ML within the RRM. As the packet scheduler within RRM aims to dynamically allocate the available resources between the mobile users at each TTI, the literature sees many scheduling schemes proposed to deal with different QoS provisioning strategies for a wide range of applications. In the frequency domain, users compete for radio resources according to some scheduling rules that focus mainly on specific QoS objectives. For example, the Proportional-Fair (PF) [50] scheduler provides a trade-off between user fairness and throughput maximization. For real-time applications, the EXPonential (EXP) rule deals with delay minimization [51]. Other frequency-based schedulers such as the Opportunistic Packet Loss Fair (OPLF) [52] and the Barrier Function (BF) [53] are designed to deal with Packet Loss Rate (PLR) minimization and with meeting the Guaranteed Bit Rate (GBR) requirements, respectively.

Frequency domain schedulers are used in conjunction with time domain schemes to attain a certain prioritization between traffic classes with heterogeneous QoS requirements. Most of these hybrid schedulers will always favor users with more stringent QoS requirements to be scheduled in the frequency domain [54–57]. Moreover, the QoS provisioning scheme is divided among time and frequency domains. For example, the scheduler in [54] pre-selects users with the highest head-of-line packet delay and the frequency domain focuses more on meeting the GBR requirement for each preselected user. The Required Activity Detection Scheduler (RADS) [56] deals with delay minimization in the time domain, whereas the frequency domain performs the PF scheduling rule to achieve proper throughput-fairness trade-offs. To minimize PLR and packet delay, the Frame Level Scheduler (FLS) [57] estimates in the time domain the amount of real-time data to be transmitted in the next frame, while the same PF scheduling rule is performed in the frequency domain. In the frequency domain, other schedulers use heuristic algorithms to minimize the throughput loss that can be caused by the time domain prioritization [58; 59]. In multi-user, multi-service and multi-network environments, these decoupled time-frequency schedulers could be employed together with network reputation algorithms [60] to select the most convenient network that enables QoS provisioning.

Due to application diversity and the heterogeneity of QoS requirements, most of these state-of-the-art schedulers are rather static, being unable to meet the QoS requirements under dynamic networking conditions [61]. By performing time domain prioritization under the same static metric, some traffic classes will be over-provisioned while others will be starved, resulting in poor QoS provisioning. Moreover, only certain QoS objectives are targeted in each scheduling domain while the remaining ones are ignored. To improve the scheduling performance and significantly increase the fraction of time (in TTIs) when the heterogeneous QoS constraints are met, Reinforcement Learning (RL) is seen as a promising solution [32; 62] that will learn the most convenient scheduling strategy to be applied in each scheduler state.

In resource scheduling problems [35], different RL algorithms are studied to get the best policy for on-line parameterization of the PF scheduling rule to meet a given fairness criterion. The RL solution proposed in [63] selects at each TTI in the frequency domain the most suitable scheduling rule to maximize the multi-objective QoS target in terms of GBR, delay and PLR for homogeneous traffic only. Similarly, a RL framework is proposed in [64] for homogeneous traffic with constant and variable bit rates, to deal with delay and PLR objectives. The work in [64] provides a comparison between different RL algorithms and aims to find the most suitable scheduling policy for various time window lengths used to compute the online PLR indicators. Compared to such approaches where only homogeneous traffic is considered, the work in [49] proposes an intelligent 5G downlink scheduling framework aiming at improving QoS provisioning in terms of GBR, delay and PLR requirements for heterogeneous traffic. However, to deal with heterogeneous traffic scheduling, a two-stage learning approach is proposed in [65]. The first stage employs a separate learning phase for each traffic class to approximate the best scheduling rule to be used. The second stage decides the best prioritization order of the traffic classes at each TTI. Only when both learning stages are completed, the entire structure can be exploited. Compared to [65], in [49], the proposed framework will learn both decisions at once, involving a much lower system complexity.

4.4.6.2 Network Slicing

The **network slicing** concept introduces the idea of splitting a shared physical network infrastructure into multiple logical networks, referred to as slices. These slices are controlled and managed independently by the slice owners (i.e., Over-The-Top (OTT) service providers or Virtual Mobile Network Operators (VMNO)) and can be used by one or multiple tenants (i.e., users). Each slice is allocated a set of network functionalities that are used only by that slice (the isolation concept) and which are selected from the shared network infrastructure. For example, one slice will be dedicated to collecting real-time data traffic for URLLC, while the other will be devoted to infotainment applications (i.e., Internet access) for eMBB as illustrated in Figure 4.14. However, given the dynamism and scalability that slicing brings, managing and orchestrating these slices through a slice orchestrator is not straightforward. Various approaches have been proposed in the literature to cope with these challenges. Kuklinski et al. [66] proposed DASMO, a distributed autonomous slice management and orchestration framework that addresses the management scalability problem. It uses the ETSI management and orchestration framework (MANO) combined with in-slice management approach to efficiently manage and orchestrate systems with a number of slices. Oladejo et al. [67] proposed a mathematical model to efficiently allocate radio resources in a two-level hierarchical network considering transmit power and allocated bandwidth constraints. The model is based on prioritizing network slices to cater for different users in a multi-tenancy network. The priority of each slice is determined by VMNOs based on which resources are allocated to the different slice while guaranteeing the minimum requirements per slice to ensure user satisfaction.

3GPP TR 28.801 [68] defines a network slice as a collection of network functions and supporting network resources that are arranged and configured to form a complete logical network that meets specific QoS characteristics required by an end-user service instance. The network slice life cycle can be divided into several stages [69], such as:

- slice preparation – where the key requirements of the network slice are identified, such as security, isolation, connectivity, QoS, reliability, etc.

- slice deployment – where the network slice is instantiated and activated for use.

- slice management – deals with the performance analysis of slice instances and the management of the required service expectations.

- slice decommissioning – the slice is deactivated, and the slice resources are freed as the slice is no longer in use.

The network slicing concept can be applied to parts of the network infrastructure, such as RAN or core only, or as an end to end solution. However, one of the most commonly studied type of network slicing is *RAN-based network slicing* which involves virtualization of the radio resources for allocation and admission control. Moreover, radio resources are assigned according to the desired QoS requirements of individual slices by sub channel allocation [70] in a shared infrastructure. These radio resources are most commonly split utilizing integer linear programming [71], or ML-based algorithms [72] as well as some heuristic approaches. However, similarly to 5G radio resource management, because of diverse QoS demands and network resources limitations, allocating resources to individual slices is a challenge in itself. Moreover, the network slicing efficiency and flexibility in service provision is enabled by the integration of SDN and NFV.

4.4.6.3 Software-Defined Networking

The standards in the traditional networking industry have been dominated by vendors with their proprietary management and proprietary solutions that sometimes fail to satisfy their customers' needs. The idea behind SDN comes to separate the control plane from the data plane giving the network administrators a more fine grained control over the traffic flows. As the technology is advancing at fast pace SDN promises to enable significant innovations within the networking paradigm. Moreover, the market begun to take shape as vendors and network operators started looking closer at the adoption and integration of OpenFlow protocol [73], which represents a promising SDN solution. OpenFlow was designed and implemented in 2008 at Stanford University, with the aim of better controlling the network. The new technology enables the programmability of the flow tables in switches and routers via a standardized interface, which is the OpenFlow protocol. The protocol defines a common API that connects a controller to switches. The switches handle the data-path functionality while the controller, a standard server, will handle the high-level networking decisions (e.g., routing, load balancing, failure recovery, etc.).

Making use of OpenFlow controllers, the network administrators will be able to define flows and policies. In this context, the control of network traffic flows is moved from the infrastructure (switches and routers) to administrators.

Thus, the promise of OpenFlow comes to separate the hardware from the control software layer enabling the network operators to build cheaper and easier to manage networks. A traditional Ethernet switch consists of two levels: *data path* – which represents the part dedicated to the hardware, responsible for packet forwarding and the *control path* – which represents the part dedicated to the software, responsible for taking decisions, similar to an operating system. In order to provide more control over the network, the OpenFlow enabled switch separates the control path from the data path. The control path is moved outside the switch enabling remotely control of the data path through a secure channel. The control functions will reside on the OpenFlow Controller, making them independent of the hardware they control.

OpenFlow provides an abstraction of the data plane through the use of flow table that can be controlled over the secure channel by the OpenFlow Controller. The flow table represents a set of entries, with each entry having three fields: a header, a counter, and an action. These entries represent rules that describe how to move a packet, appertaining to a certain flow, from the source to destination in an efficient way. When a packet is received by the OpenFlow switch, it is compared against the entries in the flow table. If a match is found in the packet header fields, the specified action is taken (e.g., forwarding, dropping, flooding, etc.). If a packet from a new flow arrives, there will be no match in the header field and the packet will be send to the controller which based on the routing algorithm will install new rules. The following packets from the same flow will be forwarded based on the previously installed rule without being send to the controller. Thus the role of the controller is to install /remove rules in the flow table, monitor the switches for traffic statistics, and respond to network events (e.g, link failure) [74]. New control functions or applications can be written on top of the OpenFlow Controller platform through its open API. This avoids the necessity to implement them in each vendor's hardware. However because OpenFlow introduces the centralized approach by having a single server handling the network-control functions, it is raising the scalability concern.

The Open Networking Foundation (ONF) was founded by companies like IBM, Cisco, Google, Facebook, Microsoft, Deutsche Telekom, etc in March 2011. The aim of ONF is to speed up innovation through software and to promote the OpenFlow technology [75]. Being part of the ONF, Google was one of the firsts to adopt SDN technology by running its entire internal network on OpenFlow. Google's networking architecture consists of two Wide Area Network backbones: (1) I-Scale which is the public Internet-facing backbone used for data delivery from Google's data centers to the end-user and (2) G-Scale which is the internal backbone that inter-connects Google's data centers worldwide. In order to start experimenting with SDN, Google used the G-Scale backbone for the OpenFlow deployment.

On the other side, within the research community, more and more OpenFlow deployments are expanding across the universities' campus with several large-scale projects like: GENI [76], FEDERICA [77], AKARI [78], etc. All these networks are used for measurements of the user-generated traffic and performance analysis of different aspects of SDN (e.g., application differentiation, predictability, scalability, load balancing, robustness, etc.).

Different networking vendors like Brocade, Cisco, Juniper or Aruba have brought the realization of SDN controller towards the commercial level through SDN/Openflow hardware release on the market [74]. However, some networking companies like Brocade and Extreme Networks leverage on the OpenDaylight [79] framework for their SDN controller solutions. A very widely used controller, especially by the research community is Floodlight controller that is originally based on Beacon controller from Stanford university and it provides simpler programming interfaces with various code samples. On the other hand, several hardware OpenFlow-enabled switches were also introduced on the market, such as: Hewlett Packard presented a high performance switch to be used within the SDN infrastructure while network companies like Brocade, Extreme Networks, Dell and IBM came out as well with their own OpenFlow-based vendor-specific SDN switches [80; 81].

4.4.6.4 Network Function Virtualization

Another key enabling technology for network slicing and 5G that works hand in hand with SDN is the NFV [82]. The concept behind NFV is to decouple various network functions like routers, firewall, catching, etc. from the dedicated specialized hardware and implement them in software as Virtual Network Function (VNF)s [83]. This will enable the VNF to be deployed on virtual machines or containers of generic hardware, such as commercial off-the-shelf servers. An architectural framework for NFV has been proposed by ETSI [84–86] and it mainly consists of three key elements:

- **VNF** – representing the software implementation of the network functions that are formerly carried out by specialized hardware;

- **Network Function Virtualization Infrastructure (NFVI)** – representing the physical infrastructure including the compute, storage and network resources as well as the virtual resource exploited by the VNFs for deployment, management and execution;

- **NFV Management and Orchestration (MANO)** – deals with the life-cycle management of VNFs and the resource management of NFVI through the use of VNF managers and Virtualized Infrastructure Manager (VIM)s, respectively. Thus, with regards to VNF, MANO will control its life-cycle like starting point, termination, updates, etc. While for NFVI, MANO will manage its resources.

It is obvious that NFV could bring several potential benefits if implemented efficiently, such as [87]: reduced Capital expenditures (CapEx) and Operating Expenses (OpEx), innovate and roll out new services that best meet the needs of customers at a fast pace, enables interoperability, enables scalability, promotes an open ecosystem, etc.

In the context of 5G networks, SDN and NFV [88] provide efficiency and flexibility by decoupling the control and data plane as well as by decoupling network functions from the dedicated hardware. SDN ensures improvement in data forwarding efficacy and NFV ensures effective deployment of dedicated network functions in the core network. However, deployment and management of network functions in service-based core network represents a significant challenge as the components are connected to each other via service-based interfaces [89]. Common practice is to make use of VNFs to map the placement of virtual nodes and choose the links chaining the VNFs. Complex network theory, graph theory [90] and mathematical models are commonly used to tackle the VNF placement problem. However, most existing solutions are limited in a way that they only perform best in assumed and specific scenarios, which makes them unsuitable in case of real time or dynamic deployment of network slices.

4.4.6.5 Multi-Access Edge Computing

The current networking environment is suffering a dramatic change as new and more complex technologies are taking over. For example, the rapid growth in cloud computing and the demands for massive datacenters, increase the need for more intelligent and efficient management systems. In order to cope with these demands, the operators have started pointing their attention towards MEC. The concept behind MEC is to move away from the remote clouds/servers by bringing the computing/storage tasks closer to the edge of the network. This could enable a small-scale cloud computing environment at the gNBs or eNBs reducing the end-user latency and avoiding traffic overhead over the network and alleviating the bandwidth consumption.

Driven by the benefits of MEC, the recent literature explored the fusion of MEC with network slicing techniques [91; 92] by trying to address possible compatibility issues. In the context of 5G, the integration of MEC could help meet the KPI requirements for specific service type classes, such as URLLC where low network latency is imperative or provide compute resources to the mMTC class [93].

4.4.7   Sixth Generation – 6G

The advances of technology in the area of MEC, AI and ML, SDN as well as the rapid deployment of 5G networks are expected to revolutionize the way we communicate, perceive and compute data by enabling a ubiquitous and pervasive paradigm ecosystem [94]. However, even if the roll-out of 5G is still being carried out around the world, works are carried out towards defining the next Sixth Generation (6G) network [95]. The vision of 6G promises a highly intelligent digital society that will be extensively data centric and enabled by unlimited wireless connectivity. However, to achieve this vision it requires new performance targets that could be met through a two stage evolution approach, such as evolution from 5G to Beyond 5G as an initial step, while the second step sees the evolution towards the revolutionary 6G [96]. Table 4.4 [96] presents a comparison of the key performance indicators of 5G, Beyond 5G and future 6G.

6G envisions interactions between three worlds: the human world (e.g., senses, bodies, intelligence, etc.), the digital world (information, communication, computing, etc.) and the physical world (objects, organisms, processes, etc.) [97]. As can be noted in Table 4.4, 6G envisions network speed of 100 to 1000 times faster than that of 5G for accommodating new service classes like [96]:

- **Mobile Broadband Reliable Low Latency Communication (MBRLLC)** service class covers multimedia-rich applications like AR, VR, holographic meetings, etc. that have stringent QoS requirements like high transmission rates, low latency and high reliability. Thus, this new service class sees the combination of eMBB and URLLC to enable meeting the rate, reliability and latency requirements.

- **massive Ultra-Reliable and Low Latency Communications (mURLLC)** service class sees the fusion of the URLLC with the legacy mMTC to bring forth a reliability-latency-scalability trade-off required for

TABLE 4.4

Requirements 5G vs. Beyond 5G vs. 6G [96]

	5G	Beyond 5G	6G
Application Types	• eMBB • URLLC • mMTC	• reliable eMBB • URLLC • mMTC • Hybrid (URLLC + eMBB)	• MBRLLC • mURLLC • HCS • MPS
Device Types	• Smartphones • Sensors • Drones	• Smartphones • Sensors • Drones • XR equipment	• MBRLLC • mURLLC • HCS • MPS
Spectral and energy efficiency gains with respect to today's network	10x in bps/Hz/m^2/Joules	100x in bps/Hz/m^2/Joules	1000x in bps/Hz/m^3/Joules (volumetric)
Rate requirements	1Gbps	100 Gbps	1Tbps
End-to-end delay requirements	5ms	1ms	< 1ms
Radio-only delay requirements	100ns	100ns	10ns
Processing delay	100ns	50ns	10ns
End-to-end reliability requirements	99.999 percent	99.9999 percent	99.99999 percent
Frequency bands	• Sub-6 GHz • mmWave for fixed access	• Sub-6 GHz • mmWave for fixed access	• Sub-6 GHz • mmWave for mobile access • Exploration of higher frequency and THz bands (above 300 GHz) • Non-RF (e.g., optical, Visible Light Communication (VLC), etc.)
Architecture	• Dense sub-6 GHz small base stations with umbrella macro base stations • mmWave small cells of about 100m (for fixed access)	• Dense sub-6 GHz small cells with umbrella macro base stations • < 100m tiny and dense mmWave cells	• cell-free smart surfaces at high frequency supported by mmWave tiny cells for mobile and fixed access • temporary hotspots served by drone-carried base stations or tethered balloons • trials of tiny THz cells.

specific applications with high connection density, such as Industry 4.0-based scenarios, large-scale IIoT/IoE, etc.

• **Human-Centric Services (HCS)** service class covers wireless Brain Computer Interface (BCI) types of application that require a new performance metric defined as Quality of Physical Experience (QoPE) that merges the human factors with QoS and QoE. Thus, brain cognition, body physiology, gestures, are factors that might impact QoPE.

- **Multi-Purpose 3CLS and Energy Services (MPS)** service class relates to the delivery of Convergence of Communications, Computing, Control, Localization and Sensing (3CLS) services and their derivatives that are important for connected robotics and autonomous systems type of applications that must meet strict performance targets like control (e.g., stability), computing (e.g., computing latency), energy (e.g., target energy to transfer), localization (e.g., localization precision) and sensing and mapping functions (e.g., accuracy of a mapped radio environment).

ABI Research exposes three types of pillars that are essential for the success of 6G networks, as illustrated in Figure 4.15 [98]. These pillars are based on three types of convergence, such as: (1) *communication and computing convergence* – to enable the 6G native intelligence; (2) *telecommunications convergence* that sees the fusion of various types of networks to enable ubiquitous coverage and global communications; (3) *convergence of digital and physical spaces* for enabling the metaverse applications that seek to fuse together the virtual world and the physical space through a single technology ecosystem.

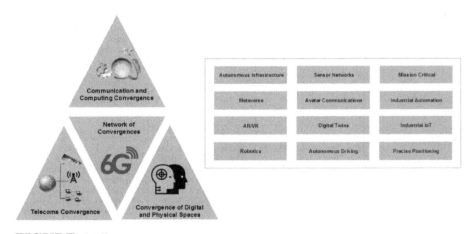

FIGURE 4.15
Three Pillars behind the 6G Vision

Thus, the advent of 6G networks sees the extension of the end-users experience beyond the physical reality as the convergence of the new and advanced technologies like digital twins, AI and ML, sensor networks, virtualization, AR/VR, computing, etc. aim to blur the boundaries between the virtual and real worlds. According to ABI Research we should expect the initial commercial deployment for 6G as early as 2028 and 2029, with the first standard technology expected around 2026.

4.5 Practical Use-Case Scenario: Network Planning for Urban Scenarios Using LTE with Altair WinProp

The aim of this practical use-case scenario is to use WinProp in order to perform network planning for an urban city area using LTE. To be able to achieve this, WallMan can be used first to generate a database map containing the geometry of a selected urban area while AMan can be used to produce different antenna pattern files. Finally, the geometry from WallMan and the antenna pattern from AMan are ready to be used by the main radio propagation simulation tool ProMan. Figure 4.16 illustrates the urban city area that we are going to use for the network planning scenario.

The overall workflow in ProMan is as follow:

- load geometry database of any urban city area that was prepared in WallMan.

- define air interface.

- define source locations.

- specify antenna pattern that was prepared in AMan.

- set up simulation, including selection of a propagation method.

- run simulation.

- inspect results.

FIGURE 4.16
WallMan – Urban City Area

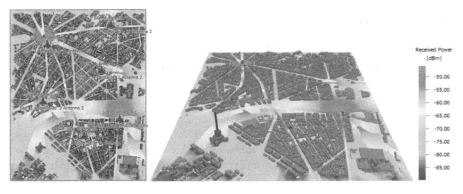

FIGURE 4.17
Received Power in 2D View and 3D View

Given the urban city area presented in Figure 4.16, we are going to create a new project in ProMan and this time we are going to use the LTE air interface made available by Altair [99] which defines OFDMA/SC-FDMA as multiple access method and FDD is used to achieve a 190 MHz separation between the uplink and the downlink. Thus, this is done by selecting File > New Project, choose *Network Planning based on description file for air interface* and select the *.wst* file defining the LTE air interface. From Scenario context menu, select *Urban Scenarios* and from databases browse and select the *.odb* file containing your urban city geometry database. The antenna pattern used is the one from a Kathrein cellular antenna, model number 741984 with the following characteristics: multi-band panel (1710-2170 MHz), dual polarization (X), half-power beam width (90°), excellent sidelobe suppression (20 dB), The antenna has a directional pattern of 120° coverage. Thus, three antennas will be used per site to offer a three sector coverage of a cell. Six sites in total are deployed over the urban city area to enable best coverage as illustrated in Figure 4.17. The antennas are positioned at a height of 35 m above ground level and the transmitter power was set to 46 dBm operating within the 2.6 GHz frequency band. The propagation model used is DPM as it takes into consideration the most relevant path making it computational effective and it is well-suited for urban environments. A static simulation environment with homogeneous traffic per cell was also considered.

Figure 4.17 illustrates the propagation results with the predicted received power from each transmitting antenna, at every location over the considered urban city area.

Figure 4.18 indicates the results for Best Server as well as the Interference and Noise results. The best server is predicted at every location over the considered urban city area and it indicates the best carrier in that particular location from which the received power is the highest. It can be noted that there are six sites deployed with three antennas each and seven carriers are used. This means that multiple antennas might have to use the same

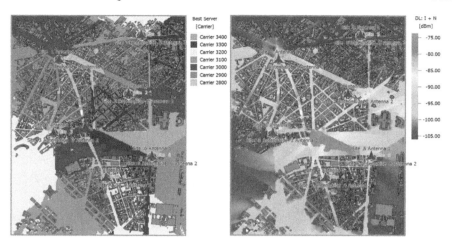

FIGURE 4.18
Best Server and Interference and Noise Results

carrier. Consequently this leads to interference as illustrated in Figure 4.18. The interference level can be further decreased by changing the carrier assignment so that antennas in close proximity could use different carriers. Figure 4.19 illustrates the maximum data rate and maximum throughput that could be achieved on the downlink at a specific location. It can be noted that the throughput exceeds the data rate by a factor of 50. This is because there are 50 resource blocks. In this use-case scenario the LTE bandwidth is set to 10 MB while the fast Fourier transform (FFT) order is 1024 indicating the maximum number of subcarriers, out of which 423 subcarriers are guard carriers and one is Direct Current (DC) carrier representing a subcarrier that has no information sent on it. Thus, a total of 600 signal carriers remain and we know that in LTE one resource block has 12 subcarriers. This means that we have a total of 50 resource blocks available.

4.6 Practical Use-Case Scenario: 5G Network Planning with Altair WinProp

The aim of this practical use-case scenario is to use WinProp in order to perform network planning for an urban city area using 5G. To be able to achieve this, WallMan can be used first to generate a database map containing the geometry of a selected urban area while AMan can be used to produce different antenna pattern files. Finally, the geometry from WallMan and the antenna pattern from AMan are ready to be used by the main radio propagation

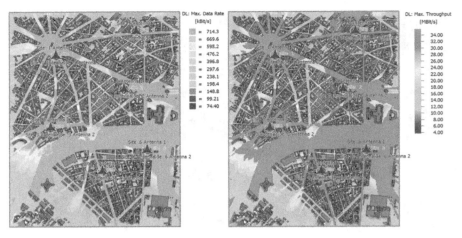

FIGURE 4.19
Downlink Data Rate and Throughput Results

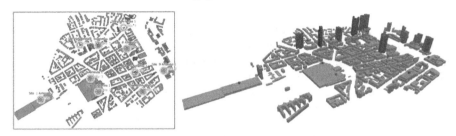

FIGURE 4.20
WallMan – Urban City Area

simulation tool ProMan. Figure 4.20 illustrates the urban city area that we are going to use for the network planning scenario.

The overall workflow in ProMan is as follow:

- load geometry database of any urban city area that was prepared in Wall-Man.

- define air interface.

- define source locations.

- specify antenna pattern that was prepared in AMan.

- set up simulation, including selection of a propagation method.

- run simulation.

- inspect results.

FIGURE 4.21
Received Power in 2D View and 3D View

Given the urban city area presented in Figure 4.20, we are going to create a new project in ProMan and this time we are going to use the 5G air interface made available by Altair [99] which defines OFDMA/SC-FDMA as multiple access method, that implements a scalable OFDM numerology with subcarrier spacing of $2^\mu \times 15\ KHz$ instead of a fixed 15 KHZ as in LTE, FDD is also used, and bandwidths options from 5 to 400 MHz. Thus, this is done by selecting File > New Project, choose *Network Planning based on description file for air interface* and select the *.wst file defining the 5G air interface. From Scenario context menu, select *Urban Scenarios* and from databases browse and select the *.odb file containing your urban city geometry database. For the antenna pattern we are going to use an omni-directional site. Thus, one antenna will be used per site. Ten sites in total are deployed over the urban city area to enable best coverage as illustrated in Figure 4.21. The antennas are positioned at a height of 15m up in the mast, with the mast positioned on the building rooftop. The transmitter power was set to 40Watts operating within the 2.1 GHz frequency band. The propagation model used is DPM as it takes into consideration the most relevant path making it computational effective and it is well-suited for urban environments. A static simulation environment with homogeneous traffic per cell was also considered.

Figure 4.21 illustrates the propagation results with the predicted received power from each transmitting antenna, at every location over the considered urban city area.

Figure 4.22 indicates the results for Best Server as well as the Interference and Noise results. The best server is predicted at every location over the considered urban city area and it indicates the best carrier in that particular location from which the received power is the highest. It can be noted that there are ten sites deployed with one antenna each and six carriers are used. This means that multiple antennas might have to use the same carrier. Consequently this leads to interference as illustrated in Figure 4.22. The interference level can be further decreased by changing the carrier assignment so that antennas in close proximity could use different carriers. Figure 4.23 illustrates the maximum data rate and maximum throughput that could be achieved on the downlink at a specific location. It can be noted that the throughput exceeds the data rate by a factor of 274. This is because there are 274 resource blocks. In this use-case scenario the 5G bandwidth is set to 100MB and numerology 1 while the fast Fourier transform (FFT) order is 4096 indicating the maximum number of subcarriers, out of which 3276 are signal carriers

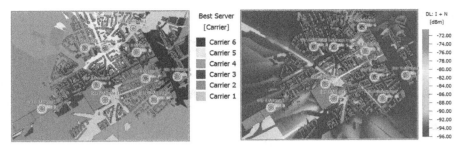

FIGURE 4.22
Best Server and Interference and Noise Results

with one resource block having 12 subcarriers. This means that we have a total of 273 resource blocks available.

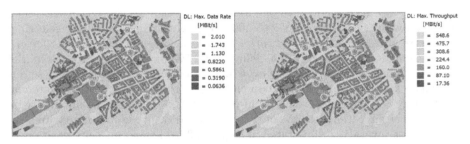

FIGURE 4.23
Downlink Data Rate and Throughput Results

Bibliography

[1] Jochen H Schiller. *Mobile communications*. Pearson education, 2003.

[2] Behrouz A. Forouzan. *Data communications and networking*. McGraw-Hill, Inc., USA, 3 edition, 2003.

[3] Stallings William. *Wireless communications and networks*. Pearson Prentice Hall, 2005.

[4] Kaveh Pahlavan and Prashant Krishnamurthy. *Principles of wireless access and localization*. John Wiley & Sons, 2013.

[5] Cory Beard and William Stallings. *Wireless communication networks and systems*. Pearson, 2015.

[6] Kaveh Pahlavan and Prashant Krishnamurthy. *Networking fundamentals: Wide, local and personal area communications.* John Wiley & Sons, 2009.

[7] Ian F. Akyildiz. Mobility management in next generation wireless systems. In *ICCCN.* IEEE Computer Society, 1998.

[8] Jiang (Linda) Xie. *Mobility Management in Next Generation All-IP Based Wireless Systems.* PhD thesis, Georgia Institute of Technology, Atlanta, GA, USA, 2004. base-search.net (ftgeorgiatech:oai: smartech.gatech.edu:1853/5190).

[9] Ian F. Akyildiz, Jiang (Linda) Xie, and Shantidev Mohanty. A survey of mobility management in next-generation all-ip-based wireless systems. *IEEE Wirel. Commun.*, 11(4):16–28, 2004.

[10] C. Perkins. IP Mobility Support for IPv4. Request for Comments (Proposed Standard) 3344, Internet Engineering Task Force, August 2002.

[11] Shengling Wang, Yong Cui, and Sajal K. Das. Intelligent mobility support for ipv6. In *LCN*, pages 403–410. IEEE Computer Society, 2008.

[12] Sri Gundavelli, Kent Leung, Vijay Devarapalli, Kuntal Chowdhury, and Basavaraj Patil. Proxy mobile ipv6. *RFC*, 5213:1–92, August 2008.

[13] Ramona Trestian. *User-centric power-friendly quality-based network selection strategy for heterogeneous wireless environments.* PhD thesis, Dublin City University, 2012.

[14] Riaz Inayat, Reiji Aibara, and Kouji Nishimura. Handoff management for mobile devices in hybrid wireless data networks. *J. Commun. Networks*, 7(1):76–86, 2005.

[15] Cong Shen and Mihaela van der Schaar. A learning approach to frequent handover mitigations in 3gpp mobility protocols. In *WCNC*, pages 1–6. IEEE, 2017.

[16] Hakim Badis and Khaldoun Al Agha. Efficient vertical handoffs in wireless overlay networks. *Comput. Artif. Intell.*, 24(5):481–494, 2005.

[17] Mehmet Fatih Tuysuz and Ramona Trestian. Energy-efficient vertical handover parameters, classification and solutions over wireless heterogeneous networks: A comprehensive survey. *Wirel. Pers. Commun.*, 97(1):1155–1184, November 2017.

[18] Ramona Trestian and Gabriel-Miro Muntean. Solutions for improving rich media streaming quality in heterogeneous network environments. 2021.

[19] M. F. Tuysuz and R. Trestian. A roadmap for a green interface selection standardization over wireless hetnets. In *2015 IEEE Globecom Workshops (GC Wkshps)*, pages 1–6, 2015.

[20] T. Bi, R. Trestian, and G. Muntean. Rload: Reputation-based load-balancing network selection strategy for heterogeneous wireless environments. In *2013 21st IEEE International Conference on Network Protocols (ICNP)*, pages 1–3, 2013.

[21] T. Bi, Z. Yuan, R. Trestian, and G. Muntean. Uran: Utility-based reputation-oriented access network selection strategy for hetnets. In *2015 IEEE International Symposium on Broadband Multimedia Systems and Broadcasting*, pages 1–6, 2015.

[22] T. Bi, R. Trestian, and G. Muntean. Reputation-based network selection solution for improved video delivery quality in heterogeneous wireless network environments. In *2013 IEEE International Symposium on Broadband Multimedia Systems and Broadcasting (BMSB)*, pages 1–8, 2013.

[23] Ramona Trestian, Olga Ormond, and Gabriel-Miro Muntean. Game theory-based network selection: Solutions and challenges. *IEEE Communications Surveys Tutorials*, 14(4):1212–1231, 2012.

[24] Chung-Hsing Yeh. A problem-based selection of multi-attribute decision-making methods. *International Transactions in Operational Research*, 9(2):169–181, 2002.

[25] Qingyang Song and Abbas Jamalipour. A network selection mechanism for next generation networks. In *IEEE International Conference on Communications, 2005. ICC 2005. 2005*, volume 2, pages 1418–1422. IEEE, 2005.

[26] Petros Nicopolitidis, Mohammed S Obaidat, Georgios I Papadimitriou, and Andreas S Pomportsis. *Wireless networks*. John Wiley & Sons, 2003.

[27] Martin Sauter. *Beyond 3G-Bringing networks, terminals and the web together: LTE, WiMAX, IMS, 4G Devices and the Mobile Web 2.0*. John Wiley & Sons, 2011.

[28] William Webb. *Wireless communications: The future*. John Wiley & Sons, 2007.

[29] John M Meredith. Technical specifications and technical reports for a utran-based 3gpp system. Technical report, 3GPP TR 21.101. 3GPP, Valbonne-FRANCE, 2011.

[30] 3rd Generation Partnership Project. High speed downlink packet access (hsdpa), overall description, stage 2 (release 7). *3GPP TS 25.308*, 2006.

[31] Yuxiang Li, Ji Fang, Kun Tan, Jiansong Zhang, Qimei Cui, and Xiaofeng Tao. Soft-lte: A software radio implementation of 3gpp long term evolution based on sora platform. *Demo in ACM MobiCom, 2009*, 2009.

[32] Ioan-Sorin Comşa. *Sustainable scheduling policies for radio access networks based on LTE technology.* PhD thesis, University of Bedfordshire, 2014.

[33] Ioan Sorin Comsa, Mehmet Aydin, Sijing Zhang, Pierre Kuonen, and Jean-Frédéric Wagen. Multi objective resource scheduling in lte networks using reinforcement learning. *International Journal of Distributed Systems and Technologies (IJDST)*, 3(2):39–57, 2012.

[34] Ioan-Sorin Comşa, Sijing Zhang, Mehmet Aydin, Pierre Kuonen, Ramona Trestian, and Gheorghiţă Ghinea. Enhancing user fairness in ofdma radio access networks through machine learning. In *2019 Wireless Days (WD)*, pages 1–8. IEEE, 2019.

[35] Ioan Sorin Comşa, Sijing Zhang, Mehmet Aydin, Jianping Chen, Pierre Kuonen, and Jean-Frederic Wagen. Adaptive proportional fair parameterization based lte scheduling using continuous actor-critic reinforcement learning. In *2014 IEEE global communications conference*, pages 4387–4393. IEEE, 2014.

[36] Ioan Sorin Comsa, Sijing Zhang, Mehmet Aydin, Pierre Kuonen, and Jean-Frederic Wagen. A novel dynamic q-learning-based scheduler technique for lte-advanced technologies using neural networks. In *37th Annual IEEE Conference on Local Computer Networks*, pages 332–335. IEEE, 2012.

[37] Ioan S Comşa, Mehmet Aydin, Sijing Zhang, Pierre Kuonen, and Jean-Frédéric Wagen. Reinforcement learning based radio resource scheduling in lte-advanced. In *The 17th International Conference on Automation and Computing*, pages 219–224. IEEE, 2011.

[38] Ioan-Sorin Comşa, Sijing Zhang, Mehmet Aydin, Pierre Kuonen, Ramona Trestian, and Gheorghiţă Ghinea. A comparison of reinforcement learning algorithms in fairness-oriented ofdma schedulers. *Information*, 10(10):315, 2019.

[39] Erik Dahlman, Stefan Parkvall, Johan Skold, and Per Beming. *3G evolution: HSPA and LTE for mobile broadband.* Academic press, 2010.

[40] Sami Tabbane. *Traffic engineering and advanced wireless network planning.* ITU Asia-Pacific Centre of Excellence Training, Suva, Fiji, 2018.

[41] VNI Cisco et al. Cisco visual networking index: Forecast and trends 2018-2023 white paper. 2019.

[42] Ramona Trestian, Ioan-Sorin Comsa, and Mehmet Fatih Tuysuz. Seamless multimedia delivery within a heterogeneous wireless networks environment: Are we there yet? *IEEE Communications Surveys Tutorials*, 20(2):945–977, 2018.

[43] ChienHsiang Wu and Chin-Feng Lai. A survey on improving the wireless communication with adaptive antenna selection by intelligent method. *Computer Communications*, 181:374–403, 2022.

[44] Mohamed Benzaghta and Khaled M Rabie. Massive mimo systems for 5g: A systematic mapping study on antenna design challenges and channel estimation open issues. *IET Communications*, 15(13):1677–1690, 2021.

[45] Yong Niu, Yong Li, Depeng Jin, Li Su, and Athanasios V Vasilakos. A survey of millimeter wave communications (mmwave) for 5g: opportunities and challenges. *Wireless networks*, 21(8):2657–2676, 2015.

[46] SM Riazul Islam, Nurilla Avazov, Octavia A Dobre, and Kyung-Sup Kwak. Power-domain non-orthogonal multiple access (noma) in 5g systems: Potentials and challenges. *IEEE Communications Surveys & Tutorials*, 19(2):721–742, 2016.

[47] Malte Schellmann, Zhao Zhao, Hao Lin, Pierre Siohan, Nandana Rajatheva, Volker Luecken, and Aamir Ishaque. Fbmc-based air interface for 5g mobile: Challenges and proposed solutions. In *2014 9Th international conference on cognitive radio oriented wireless networks and communications (CROWNCOM)*, pages 102–107. IEEE, 2014.

[48] Long Qu, Jiaming He, and Chadi Assi. Understanding the benefits of successive interference cancellation in multi-rate multi-hop wireless networks. In *2014 IEEE International Conference on Communications (ICC)*, pages 354–360, 2014.

[49] Ioan-Sorin Comșa, Ramona Trestian, Gabriel-Miro Muntean, and Gheorghiță Ghinea. 5mart: A 5g smart scheduling framework for optimizing qos through reinforcement learning. *IEEE Transactions on Network and Service Management*, 17(2):1110–1124, 2019.

[50] A. Jalali, R. Padovani, and R. Pankaj. Data Throughput of CDMA-HDR a High Efficiency-High Data Rate Personal Communication Wireless System. In *2000 IEEE 51st Vehicular Technology Conference Proceedings VTC2000-Spring*, volume 3, pages 1854 – 1858, May 2000.

[51] B. Sadiq, R. Madan, and A. Sampath. Downlink Scheduling for Multiclass Traffic in LTE. *in EURASIP Journal on Wireless Communications and Networking*, 2009(14):1–18, 2009.

[52] N. Khan, M.G. Martini, Z. Bharucha, and G. Auer. Opportunistic Packet Loss Fair Scheduling for Delay-Sensitive Applications over LTE Systems. In *IEEE Wireless Communications and Networking Conference*, volume 1, pages 1456 – 1461, April 2012.

[53] M. Lundevall, B. Olin, J. Olsson, N. Wiberg, S. Wanstedt, J. Eriksson, and F. Eng. Streaming Applications Over HSDPA in Mixed Service scenarios. In *IEEE Vehicular Technology Conference VTC2004-Fall*, volume 1, pages 841 – 845, April 2005.

[54] B. Bojovic and N. Baldo. A New Channel and QoS Aware Scheduler to Enhance the Capacity of Voice over LTE Systems. In *IEEE 11th International Multi-Conference on Systems, Signals and Devices (SSD14)*, pages 1 – 6, February 2014.

[55] F.T.S. Avocanh, M. Abdennebi, and J. Ben-Othman. An Enhanced Two Level Scheduler to Increase Multimedia Services Performance in LTE Networks. In *IEEE International Conference on Communications (ICC)*, pages 2351 – 2356, June 2014.

[56] G. Monghal, D. Laselva, P.-H. Michaelsen, and J. Wigard. Dynamic Packet Scheduling for Traffic Mixes of Best Effort and VoIP Users in E-UTRAN Downlink. In *IEEE Vehicular Technology Conference (VTC-Spring)*, pages 1 – 5, May 2010.

[57] G. Piro, L.A. Grieco, G. Boggia, R. Fortuna, and P. Camarda. Two-Level Downlink Scheduling for Real-Time Multimedia Services in LTE Networks. *in IEEE Transactions on Multimedia*, 13:1052 – 1065, 2011.

[58] K. Wang, X. Li, H. Ji, and X. Zhang. Heterogeneous Traffic Scheduling in Downlink High Speed Railway LTE Systems. In *IEEE Global Communications Conference (GLOBECOM)*, pages 1452 – 1457, December 2013.

[59] W.C. Chung, C. J. Chang, and L.C. Wang. An Intelligent Priority Resource Allocation Scheme for LTE-A Downlink Systems. *in IEEE Wireless Communications Letters*, 1(3):241 – 244, 2012.

[60] C. Desogus, M. Anedda, M. Murroni, and G.-M. Muntean. A Traffic Type-Based Differentiated Reputation Algorithm for Radio Allocation During Multi-Service Content Delivery in 5G Heterogeneous Scenarios. *in IEEE Access*, 7:27720 – 27735, 2019.

[61] O. Grndalen, A. Zanella, K. Mahmood, M. Carpin, J. Rasool, and O. N. Sterb. Scheduling Policies in Time and Frequency Domains for LTE Downlink Channel: A Performance Comparison. *in IEEE Transactions on Vehicular Technology*, 66(4):3345 – 3360, April 2017.

[62] R. S. Sutton and A. G. Barto. *Reinforcement learning: An introduction.* IMT Press Cambridge, 2017.

[63] Ioan-Sorin Comsa, Antonio De-Domenico, and Dimitri Ktenas. Qos-driven scheduling in 5g radio access networks-a reinforcement learning approach. In *GLOBECOM 2017-2017 IEEE Global Communications Conference*, pages 1–7. IEEE, 2017.

[64] Ioan-Sorin Comşa, Sijing Zhang, Mehmet Emin Aydin, Pierre Kuonen, Yao Lu, Ramona Trestian, and Gheorghiţă Ghinea. Towards 5g: A reinforcement learning-based scheduling solution for data traffic management. *IEEE Transactions on Network and Service Management*, 15(4):1661–1675, 2018.

[65] I. S. Comsa, A. De Domenico, and Dimitri Ktenas. Method for Allocating Transmission Resources Using Reinforcement Learning, April 2019.

[66] Slawomir Kukliński and Lechosław Tomaszewski. Dasmo: A scalable approach to network slices management and orchestration. In *NOMS 2018-2018 IEEE/IFIP Network Operations and Management Symposium*, pages 1–6. IEEE, 2018.

[67] Sunday O Oladejo and Olabisi E Falowo. 5g network slicing: A multi-tenancy scenario. In *2017 Global Wireless Summit (GWS)*, pages 88–92. IEEE, 2017.

[68] 3GPP TR 28.801. *Study on Management and Orchestration of Network Slicing for Next Generation Network*. 3GPP Technical Report V15.1.0, 2018.

[69] Shalitha Wijethilaka and Madhusanka Liyanage. Survey on network slicing for internet of things realization in 5g networks. *IEEE Communications Surveys & Tutorials*, 23(2):957–994, 2021.

[70] Yuan Ai, Gang Qiu, Chenxi Liu, and Yaohua Sun. Joint resource allocation and admission control in sliced fog radio access networks. *China Communications*, 17(8):14–30, 2020.

[71] Menglan Jiang, Massimo Condoluci, and Toktam Mahmoodi. Network slicing management & prioritization in 5g mobile systems. In *European Wireless 2016; 22th European Wireless Conference*, pages 1–6. VDE, 2016.

[72] Jie Mei, Xianbin Wang, and Kan Zheng. An intelligent self-sustained ran slicing framework for diverse service provisioning in 5g-beyond and 6g networks. *Intelligent and Converged Networks*, 1(3):281–294, 2020.

[73] Nick McKeown, Tom Anderson, Hari Balakrishnan, Guru Parulkar, Larry Peterson, Jennifer Rexford, Scott Shenker, and Jonathan Turner. Openflow: enabling innovation in campus networks. *ACM SIGCOMM computer communication review*, 38(2):69–74, 2008.

[74] Ahmed Al-Jawad. *SPolicy-Based Network Management with End-to-End QoS Solution in Software-Defined Networking*. PhD thesis, Middlesex University, 2022.

[75] Sibylle Schaller and Dave Hood. Software-defined networking architecture standardization. *Computer standards & interfaces*, 54:197–202, 2017.

[76] Jonathan S Turner. A proposed architecture for the geni backbone platform. In *Proceedings of the 2006 ACM/IEEE symposium on Architecture for networking and communications systems*, pages 1–10, 2006.

[77] Peter Szegedi, Sergi Figuerola, Mauro Campanella, Vasilis Maglaris, and Cristina Cervelló-Pastor. With evolution for revolution: managing federica for future internet research. *IEEE Communications Magazine*, 47(7):34–39, 2009.

[78] Hiroaki Harai. Designing new-generation network-overview of akari architecture design. In *2009 Asia Communications and Photonics conference and Exhibition (ACP)*, pages 1–2. IEEE, 2009.

[79] Jan Medved, Robert Varga, Anton Tkacik, and Ken Gray. Opendaylight: Towards a model-driven sdn controller architecture. In *Proceeding of IEEE International Symposium on a World of Wireless, Mobile and Multimedia Networks 2014*, pages 1–6. IEEE, 2014.

[80] Maciej Kuźniar, Peter Perešíni, and Dejan Kostić. What you need to know about sdn flow tables. In *International Conference on Passive and Active Network Measurement*, pages 347–359. Springer, 2015.

[81] Bruno Astuto A Nunes, Marc Mendonca, Xuan-Nam Nguyen, Katia Obraczka, and Thierry Turletti. A survey of software-defined networking: Past, present, and future of programmable networks. *IEEE Communications Surveys & Tutorials*, 16(3):1617–1634, 2014.

[82] Bruno Chatras and François Frédéric Ozog. Network functions virtualization: The portability challenge. *IEEE Network*, 30(4):4–8, 2016.

[83] Yichao Jin and Yonggang Wen. When cloud media meet network function virtualization: Challenges and applications. *IEEE MultiMedia*, 24(3):72–82, 2017.

[84] NFVISG ETSI et al. Network functions virtualisation (nfv); management and orchestration. *NFV-MAN*, 1:v0, 2014.

[85] Borja Nogales, Ivan Vidal, Diego R Lopez, Juan Rodriguez, Jaime Garcia-Reinoso, and Arturo Azcorra. Design and deployment of an open management and orchestration platform for multi-site nfv experimentation. *IEEE Communications Magazine*, 57(1):20–27, 2019.

[86] Faqir Zarrar Yousaf, Vincenzo Sciancalepore, Marco Liebsch, and Xavier Costa-Perez. Manoaas: A multi-tenant nfv mano for 5g network slices. *IEEE Communications Magazine*, 57(5):103–109, 2019.

[87] William Stallings. *Foundations of modern networking: SDN, NFV, QoE, IoT, and Cloud*. Addison-Wesley Professional, 2015.

[88] GSNFV ETSI. "etsi gs nfv-eve 005 v1.1.1: Network functions virtualisation (nfv); ecosystem; report on sdn usage in nfv architectural framework." *Last Accessed: March,* 2018.

[89] Lu Ma, Xiangming Wen, Luhan Wang, Zhaoming Lu, and Raymond Knopp. An sdn/nfv based framework for management and deployment of service based 5g core network. *China Communications,* 15(10):86–98, 2018.

[90] Hsu-Tung Chien, Ying-Dar Lin, Chia-Lin Lai, and Chien-Ting Wang. End-to-end slicing as a service with computing and communication resource allocation for multi-tenant 5g systems. *IEEE Wireless Communications,* 26(5):104–112, 2019.

[91] Luca Cominardi, Thomas Deiss, Miltiadis Filippou, Vincenzo Sciancalepore, Fabio Giust, and Dario Sabella. Mec support for network slicing: Status and limitations from a standardization viewpoint. *IEEE Communications Standards Magazine,* 4(2):22–30, 2020.

[92] Adlen Ksentini and Pantelis A Frangoudis. Toward slicing-enabled multi-access edge computing in 5g. *IEEE Network,* 34(2):99–105, 2020.

[93] Yuyi Mao, Changsheng You, Jun Zhang, Kaibin Huang, and Khaled B Letaief. A survey on mobile edge computing: The communication perspective. *IEEE Communications Surveys & Tutorials,* 19(4):2322–2358, 2017.

[94] Grigorios Kakkavas, Adamantia Stamou, Vasileios Karyotis, and Symeon Papavassiliou. Network tomography for efficient monitoring in sdn-enabled 5g networks and beyond: Challenges and opportunities. *IEEE Communications Magazine,* 59(3):70–76, 2021.

[95] Ioannis Tomkos, Dimitrios Klonidis, Evangelos Pikasis, and Sergios Theodoridis. Toward the 6g network era: Opportunities and challenges. *IT Professional,* 22(1):34–38, 2020.

[96] Walid Saad, Mehdi Bennis, and Mingzhe Chen. A vision of 6g wireless systems: Applications, trends, technologies, and open research problems. *IEEE Network,* 34(3):134–142, 2019.

[97] 5G Infrastructure Association et al. European vision for the 6g network ecosystem. *V1. 0,* pages 06–07, 2021.

[98] Malik Saadi. 6G: The Network of Technology Convergence. *ABI Research, Tech. Rep.,* 2022.

[99] Altair. Altair winprop 2021.1 user guide. *Altair Engineering Inc.,* 2021.

Chapter 5

Satellite Communications

"I just have my own attitude. I'm out here to get the job done, and I knew I had the ability to do it, and that's where my focus was...My head is not in the sand. But my thing is, if I can't work with you, I will work around you. I was not about to be [so] discouraged that I'd walk away. That may be a solution for some people, but it's not mine."

Annie Easley
Computer scientist, Mathematician,
and Rocket Scientist at the Lewis Research Centre at NASA,
Laid the foundations for space shuttle launches.

5.1 The Future of Satellite Communications

We have seen previously that the current 5G networks continue to be defined as terrestrial mobile communication networks and the vision of 6G is to go a step further and make everything interconnected by enabling a global communication system that aims to provide coverage over the entire earth's surface area including oceans, deserts, forests and airspace [3]. Thus, to be able to achieve this, satellite systems are seen as a significant solution for integration within the 6G ecosystem. As we have previously seen, 6G aims to achieve this telecommunications convergence that sees the fusion of various types of networks including satellite systems that could enable access to communication services for mobile users in developing areas, emergency areas, planes, trains, ships, etc. 6G also aims to revolutionize the way we communicate, interact with machines, perceive and compute data, a scenario that together with the telecommunications convergence has the potential to unlock greater opportunities and bridge the digital divide around the world.

An example of the telecommunications convergence vision that could enable an intelligent 6G network was introduced by Rasti et al. as illustrated in Figure 5.1 [4].

As depicted in Figure 5.1 the converged telecommunications ecosystem consists of several enabling technologies at the physical layer, including [4]:

DOI: 10.1201/9781003222095-5

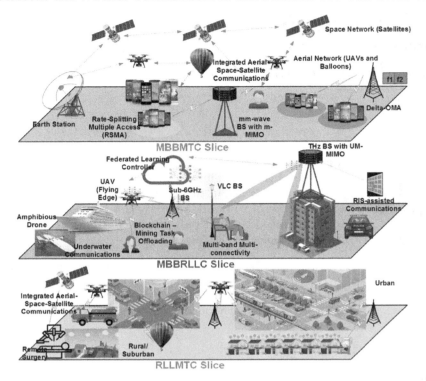

FIGURE 5.1
Intelligent 6G Network Vision

the use of terahertz (THz) and visible light communications licensed and unlicensed spectrum, integration of ultra-massive spatially modulated MIMO (UM-SM-MIMO), the use of reconfigurable intelligent surface (RIS) for assisted communications, convergence of Underwater-Terrestrial-Air-Space integrated Networks (UTASNet), adoption of dynamic network slicing and virtualization, different types of non-orthogonal multiple access (NOMA) including delta-OMA (D-OMA) and rate splitting multiple access (RSMA) [5].

Moreover, Figure 5.1 illustrates three slices each corresponding to one type of service class defined as: (1) *Mobile Broadband Reliable Low Latency Communication (MBBRLLC)* that represents the fusion of FeMBB and eURLLC classes and aims to support high broadband data rates along with reliable, and low latency communication; (2) *Mobile Broadband Machine Type Communication (MBBMTC)* represents the fusion of FeMBB and umMTC classes and aims to support high broadband data rates along with massive connectivity; (3) *Reliable Low Latency Machine Type Communication (RLLMTC)* represents the fusion of eURLLC and umMTC classes that aims to support massive connectivity, reliability, and low latency. Apart from these three service classes, a fourth service class type is proposed in [3] namely *Mobile Broadband and Reliable Low Latency Machine Type Communication (MBBRLLMTC)* representing the fusion of eURLLC, FeMBB and umMTC classes which aims to support high data rates, reliability, low latency, and massive connectivity.

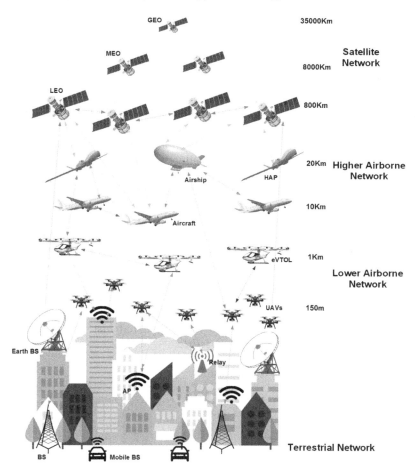

FIGURE 5.2
Future Aerial Communication Vision

Consequently, it is obvious that to be able to achieve the strict QoS requirements of various types of services and applications as well as to enable global coverage, the integration of space, terrestrial and airborne networks is mandatory [6; 7]. The vision of the future aerial communications and the integration with the terrestrial network is illustrated in Figure 5.2 [6; 7]. The vision incorporates different types of aerial technologies that could enable the communication in the sky through different connectivity platforms, such as: unmanned aerial vehicle (UAV), electric vertical take-off and landing (eVTOL), aircraft, airship, high altitude platform (HAP). The satellite network consists of three types of satellites based on their altitude, such as: Low Earth Orbit (LEO), Medium Earth Orbit (MEO) and Geostationary Earth Orbit (GEO).

5.2   Satellite Basics

A basic satellite communication network is illustrated in Figure 5.3. The system enables the communication from one point on the Earth to another regardless of their location making the satellite communication suitable for undeveloped parts of the world to enable communication. This is done through the Earth stations representing the transmitting and receiving antenna systems or it can be represented by a communication platform near Earth. The transmission from the Earth station to the satellite represents the *uplink* while *downlink* is represented by the transmission from the satellite to the Earth station. In general Frequency Division Duplex (FDD) is used for the separation of uplink and downlink frequencies. The satellite is an artificial satellite consisting of electronic equipment such as *transponder* that convert the uplink signals to downlink signals. There can be different types of transponders, such as: (1) transparent transponder that performs the basic functionality of shifting the uplink frequency to the downlink frequency, also known as *Bent Pipe* and (2) regenerative transponder that performs more advanced functions such as regenerating and formatting of the signal that lost its energy during travel [8], also known as *Processing Satellite*.

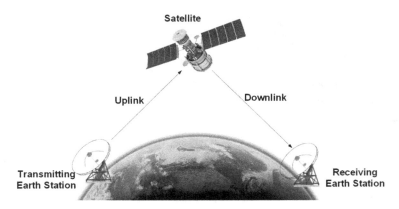

FIGURE 5.3
Basic Satellite Communication

The satellites orbit around the Earth in different planes as illustrated in Figure 5.4, such as: the equatorial orbit above the Earth's Equator, the Polar orbit that passes through both poles and the inclined orbit that refers to all the other orbits. These orbits can be either circular with their centre at Earth's centre or elliptical with one foci at Earth's center.

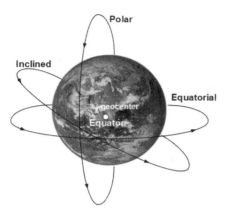

FIGURE 5.4
Satellite Orbits

The time required for a satellite to make a complete trip around the Earth is referred to as the satellite's period. Kepler's law defines the period of a satellite as a function of the distance of the satellite from the geocentre (Earth's centre), as given by the equation below [8]:

$$Period[sec] = C \times distance[km] \cdot 1.5 \qquad (5.1)$$

where C is a constant of value $\approx 1/100$. For example, knowing that the Earth's radius is 6,378 km and using Kepler's formula we can calculate that a satellite with a period of 24 h is located at 35,786 km from Earth. This period is the same as Earth's rotation which makes the satellite located at this distance from Earth seem *stationary*. Consequently the orbit is referred to as a geosynchronous orbit. Consequently, depending on their altitude, four different types of orbits can be identified, such as: *Geostationary Earth Orbit (GEO)* – including the satellites located at a distance of almost 36,000 km to the Earth, *Medium Earth Orbit (MEO)* – including the satellites operating at a distance of about 5,000 to 12,000 km, *Low Earth Orbit (LEO)* – including the satellites located at a distance of 500 – 1,500 km above Earth surface, and *Highly Elliptical Orbit (HEO)* consisting of all the satellites with non circular orbits [9]. However, there are two Van Allen belts that contain ionized (aionaized) particles located at about 2,000 – 6,000 km (inner Van Allen belt) and about 15,000 – 30,000 km (outer Van Allen belt) making the satellite communications impossible in these orbits. Consequently the MEO orbits are located between these two Van Allen belts.

In order for a satellite to maintain the same distance to Earth's surface while orbiting around the Earth, they need to follow a simple law. Consequently, the *attractive force* F_g is defined as [9]:

$$F_g = m \cdot g(R/r^2) \tag{5.2}$$

where m represents the mass of the satellite, R is the Earth's radius ($\approx$ 6,378 km), r represents the distance to the centre of the Earth and g is the acceleration of gravity ($\approx 9.81 m/s^2$).

The *centrifugal force* F_c is defined as:

$$F_c = m \cdot r \cdot w^2 \tag{5.3}$$

where w is the angular velocity ($w = 2\pi f$, with f being the rotation frequency). Thus, in order to maintain the satellite in a stable circular orbit, both forces must be equal $F_g = F_c$. Consequently, we can notice that in this case the mass of the satellite m is irrelevant and the distance r of the satellite to the Earth's centre is given by:

$$r = \sqrt[3]{\frac{gR^2}{(2\pi f)^2}} \tag{5.4}$$

Thus, this means that the distance of a satellite to the Earth's surface given by r depends on its rotation frequency f. As we have mentioned previously, when the period of the satellite equals 24 h the satellite is located exactly at 35,786 km from Earth implying geostationary satellites.

Two important parameters that need to be defined when dealing with satellite communications are: (1) *inclination angle* which represents the angle between the satellite's orbit plane and the equatorial plane. When the satellite is exactly above the equator that means that the inclination angle is zero degrees. (2) *elevation angle* which represents the angle between the centre of the satellite beam and Earth's surface. There is a minimal elevation required to be able to communicate with the satellite. This is because, depending on the elevation of a satellite, the signal will have to penetrate a smaller or larger portion of the atmosphere which in turn will attenuate the signal due to various atmospheric conditions, such as rain absorption, fog absorption and atmospheric absorption. The propagation loss L depends on the distance r between the satellite and Earth's centre and is given by [9]:

$$L = (\frac{4\pi r f}{c})^2 \tag{5.5}$$

where c is the speed of light and f the carrier frequency. This means that the received signal power decreases with the square of the distance. Thus, in general, parameters like attenuation or received power are determined by four parameters, such as: sending power, gain of the sending antenna, distance between the sender and the receiver and the gain of the receiving antenna.

TABLE 5.1

Satellite Communications Frequency Bands [1]

Band	Frequency Range	Total Bandwidth	General Application
L	1 to 2 GHz	1 GHz	Mobile satellite service (MSS)
S	2 to 4 GHz	2 GHz	MSS, NASA, deep space research
C	4 to 8 GHz	4 GHz	Fixed Satellite Service (FSS)
X	8 to 12.5 GHz	4.5 GHz	FSS military, terrestrial earth exploration, and meteorological satellites
Ku	12.5 to 18 GHz	5.5 GHz	FSS, broadcast satellite service (BSS)
K	18 to 26.5 GHz	8.5 GHz	BSS, FSS
Ka	26.5 to 40 GHz	13.5 GHz	FSS

However, because in the wireless environment we usually deal with variation in the received signal strength especially due to multi-path propagation as well as interruptions due to shadowing of the signal, in satellite system we typically need LoS to enable communication.

In terms of frequencies used for satellite communications, Table 5.1 [1] gives the band names and frequencies for each range as well as the total bandwidth and general applications. It goes without saying that as you go higher in frequency bands the total bandwidth is increasing. However, it is also known that at higher frequencies the impact of the transmission impairments greater.

A typical basic scenario of a satellite system for global mobile communication is illustrated in Figure 5.5 [9]. Depending on the type of the satellite and its distance from Earth, each satellite will have a so called *footprint* which is the coverage area on Earth where the signals from the satellite can be received (similar to a cell for cellular systems). However, some satellites can be equipped with smart antennas that can create small cells referred to as spot beams. A mobile user located within a satellite's footprint can communicate with teh satellite via a mobile user link (MUL). The satellite communicates with the Earth station (base station), that can also act as a gateway to other networks, via the gateway link (GWL). Moreover, some satellites are able to communicate directly with each other via inter-satellite link (ISL). However, not all satellites are equipped with ISL as this requires higher system complexity and processing at the satellite which might shorten their lifetime due to increase in fuel consumption. The advantages of the ISL is that it reduces the number of Earth gateways required to establish a communication which means that it will reduce the latency by keeping the communication within the satellite network as long as possible and using a minimum of uplink and downlink connections.

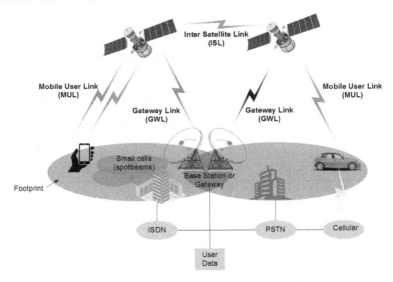

FIGURE 5.5
Classical Satellite System for Global Mobile Communications

5.2.1   GEO Satellites

The GEO satellites orbit at 35,786 km distance to the Earth surface in the equatorial plane, meaning that their inclination angle is zero degrees. Consequently, the GEO satellites perform a complete rotation within exactly one day (24h) which makes the satellite to be synchronous to the Earth rotation making them look as stationary. This means that receiver antennas do not require adjustments and they can be kept in fixed positions. The advantage of the GEO satellites is that they have a large footprint and only three GEO satellites equidistant from each other, are required to cover the entire Earth surface. However, this also means that the reuse of frequencies is not possible, they require a high transmit power and they introduce a high latency due to the long distance from Earth. These types of satellites are typically used for radio and TV transmission as they are not suitable for mobile data transmissions.

5.2.2   MEO Satellites

The MEO satellites orbit at around 5,000 to 12,000 km above Earth surface between the two Van Allen belts and it takes them approximately 6 to 8 hours for a complete rotation around the Earth. The diameter of coverage for the MEO satellites footprint is 10,000 to 15,000 km, which means that a higher number of satellites (a dozen) are required to provide worldwide coverage as compared to the GEO satellites. However, because also these satellites are located at a considerable distance from Earth, they require high transmit power and they introduce a high latency (e.g., 70–80 ms).

5.2.3 LEO Satellites

The LEO satellites orbit at around 500 – 1500 km above Earth surface and perform a complete rotation within 90 to 120 minutes. This means that the LEO satellites are moving at a higher speed (around 20,000 to 25,000 km/h) as compared to GEO and MEO making them visible from Earth for around 10 minutes. As LEO satellites are closer to Earth they require lower transmit power and have a smaller footprint with better frequency reuse. Thus, a constellation of LEO satellites is required to work together as a network for world wide coverage. This makes them a more complex system due to moving satellites and also introduces another challenge for the LEO satellite systems in terms of increased number handovers from one satellite to another. The advantage is that the latency is comparable with terrestrial long distance networks and global radio coverage is possible.

5.3 Applications of Satellites

We have seen that the advent of 6G sees the integration of satellite and terrestrial systems for global communication coverage. Both satellite and terrestrial mobile systems have developed and evolved independently across the years. While the cellular mobile systems had a spectacular and rapid evolution over the years from one generation to another, 1G to 5G currently, and the upcoming 6G, the satellite mobile systems faced challenges in commercialization. This is because the competing cellular mobile systems were developed and deployed at a higher pace than satellite mobile systems. Figure 5.6 illustrates the development period of satellite systems [10].

We can notice that during 1990s the satellite mobile systems emerged as the first climactic development with examples of GEO and LEO systems, such as Inmarsat and Iridium, respectively. Only starting with the 2010s there was a second climactic development for the satellite systems represented by the significant development of LEO systems, such as Oneweb, Starlink, Kuiper, Telesat, etc. This evolution from the first climactic development to the second climactic development, also sees the evolution of the satellite systems from mobile communications to broadband communications. This was due to the accelerated innovations in satellite manufacturing and rocket launching technologies, integrated circuit technologies, as well as communications technologies that led to significant cost reductions and thus with thousands to ten of thousands satellites, especially LEO satellites, existing in a constellation. Consequently, this led to the satellite's applications to extend from various

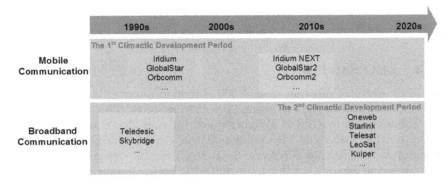

FIGURE 5.6
Satellite Development Periods

industries to public service, with different application areas within urban, rural or remote environments as illustrated in Figure 5.7 [11]. For example, the satellite communication systems can be used to reinforce the terrestrial networks and provide expanded connectivity in rural and remote areas [2]. In urban areas the satellite communication system can be used in conjunction with the terrestrial networks to provide fixed network backhaul, IoT connectivity, enabling connected cars, disaster relief, etc. Moreover, recently the satellite communications systems have been seen as important enabler for air-borne and maritime communication platforms.

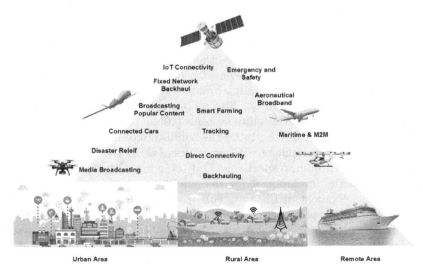

FIGURE 5.7
Applications of Satellites in the New Space Era

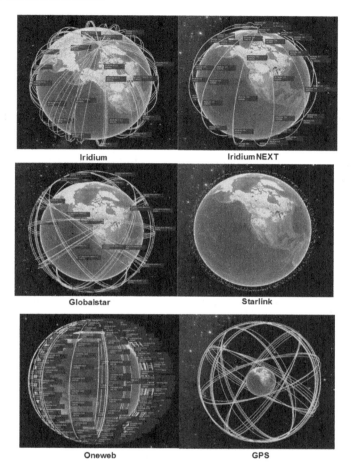

FIGURE 5.8
Examples of Satellite Systems

Some examples of satellite systems are presented in Figure 5.8[1]

5.3.1 Iridium

Iridium is a LEO-based satellite mobile communication network started by Motorola in the 1990s and it took eight years to complete [8]. The Iridium satellite system uses 66 active satellites that are placed in circular polar orbits at a distance of around 780 km [12] from the Earth surface and it offers complete Earth coverage including the polar regions, aeronautical and oceanic regions [13]. The 66 LEO satellites are divided into six orbits with 11 satellites

---

[1]Celestrak: http://www.celestrak.com/NORAD/elements/

per orbit. The satellites are equipped with on board processing capabilities to enable FDM/TDMA operations and ISL for inter satellite communications. The Iridium system operates in the L band and it was designed to support voice and low bit rate data transmissions anywhere and at any time due to their global coverage. Moreover, each satellite can employ 48 spot beams with a maximum number of active spot beams at any moment of around 2,000 [8]. This means that the system can have around 2,000 overlapping cells projected on Earth. Due to the use of ISL, most of the traffic routing is done through the satellites which decreases the latency and the number of Earth stations required. The next generation of Iridium constellation is referred to as IridiumNext and has completely replaced the original Iridium system with new and upgraded satellites. This enabled new and faster broadband services. The IridiumNext constellation consists of 81 satellites built by Thales Alenia Space, with 66 in the active constellation, 9 in-orbit spares, and 6 ground spares.[2]

5.3.2 Globalstar

The Globalstar is another LEO-based satellite mobile communication network that was started by a limited partnership with Loral and Qualcomm as main partners. The Globalstar satellite system uses 48 active satellites that are placed in eight circular orbits at a distance of 1,414 km from the Earth surface and with six satellites per orbit [12]. Unlike the Iridium satellites, the Globalstar system uses bent-pipe LEO satellites, without ISL and they can employ 16 spot beams. Thus, Globalstar system does not provide global coverage like Iridium and it enables communications only in areas within ±70 latitudes and Earth stations/gateways present, meaning that the polar regions are not served. This means that Globalstar require an increased number of Earth stations as compared to Iridium, with a coverage area of 2,000 km per Earth station. The physical layer technique used by Globalstar is DS-CDMA and it operates within the L and S bands and offers real-time voice, data, and fax [13]. However, the communication latency might be increased as there is no direct communication between satellites and the signal has to come back to the Earth station.

5.3.3 Inmarsat

Inmarsat is a GEO-based satellite communication system that was established in 1979 as International Maritime Satellite Organization (Inmarsat) and delivers broadband communication services to enterprise, maritime, and aeronautical users [13]. The Inmarsat system evolved over time and it uses 12 GEO satellites consisting of four Inmarsat-2, five Inmarsat-3 and three Inmarsat-4 to enable the provision of mobile phone, fax, and data communications to

---

[2]https://www.iridium.com/blog/2019/02/08/iridium-next-review/

TABLE 5.2
Important Satellite Systems in Planning [2]

Constellation Name	Country	No. of Satellites	Altitudes	Service Band	Expected Start Date
SpaceX - Starlink	USA	42,000	1,150 km, 550 km, 340 km	Ku, Ka, V	2020
Oneweb	UK	5,260	1,200 km	Ku	2020
Telesat	Canada	512	1,000 km	Ka	2022
Amazon - Kuiper	USA	3236	590 km, 630 km, 610 km	Ka	2021
Lynk	USA	thousands	450-500 km	spectrum from MNOs	2023
Facebook	USA	thousands	500-550	E	2021
Aerospace Sci. Corp.	China	156	1,000	Ka	2022

the whole Earth surface, except the Poles. Starting with Inmarsat-3 the satellites can employ spot beams as well as single large global beam. Similarly to Iridium and Globalstar, the Inmarsat satellites also operate in the L band

5.3.4 Starlink

A more recent LEO-based satellite system that belongs to the second climactic development period, is Starlink initialized by the U.S. based company SPaceX in 2019 with the launch of the first batch of 60 LEO satellites into the circular Earth orbit at a distance of around 430 km from the Earth surface [14]. The Starlink project was endorsed by the FCC to extend the proposed constellation to up to 42,000 Starlink LEO satellites [15]. Figure 5.8 shows the Starlink constellation with currently (as of 2022) more than 2,000 active Starlink satellites. Due to the high number of satellites, the distance between them is controlled by an autonomous collision avoidance system that makes use of input data from the department of defense's debris tracking system in order to avoid collisions with space debris and other spacecrafts [2]. The Starlink system operates in the Ka and Ku bands spectrum, with an estimated latency between from 15ms to 100ms and downlink speeds of up to 103 Mbps.

Table 5.2 offers an overview of some of the most important recent satellite systems that are currently in planning or under construction [2].

5.3.5   Global Positioning System

The Global Positioning System (GPS) is a MEO-based satellite system aimed for navigation purposes that was developed by the U.S. Department of Defense in the early 1970s. The GPS satellites orbit at a distance of around 18,000 km from the Earth surface [16]. The GPS system consists of a constellation of 24 active satellites distributed over six nearly circular orbits as illustrated in Figure 5.8 in order to provide continuous positioning and timing information for land, sea and air navigation. The GPS system is designed in such a way that there will always be four satellites visible from any point on Earth at any time [8]. The functionality of the GPS systems is based on the principle of *trilateration* which involves finding any location on Earth by knowing the distance from three GPS satellites and their positions. This is done by one-way ranging that gives the distance from at least three GPS satellites (Satellites A, B, and C in Figure 5.9 [17]) defined by the center of three spheres. The intersection of the three spheres provides two points of intersections. However, one of the two points where the spheres meet it will be so unrealistic that can be eliminated. However, a fourth satellite (Satellite D in Figure 5.9) needs to be used to adjust timing offsets [1]. Thus we need at least four GPS satellites to find the exact position of any point on Earth (longitude, latitude, and altitude). Moreover the GPS system is based on the DSSS technology to keep away from unauthorized use and all satellites can use the same frequency band. However, the operation of the GPS satellite system introduces some complexities in terms of satellite synchronization, knowing the satellite locations, atmospheric effects, etc. In order to improve the accuracy of the GPS system, a differential GPS solution could be used where a terrestrial reference point is also known [17].

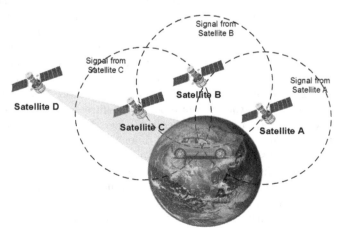

FIGURE 5.9
Global Positioning Systems

5.4 Routing and Localization

As we have seen exemplified in Figure 5.5, a typical basic satellite system consists of the satellites, which can be LEO, MEO or GEO, the Earth stations and the fixed terrestrial network part. Similarly to the cellular system, the satellite system needs to enable mobility management and localization as well as routing of data transmissions from one user to another. When dealing with GEO satellites, the routing and mobility management is similar to that of the terrestrial networks due to them being relatively static with relatively fixed beam coverage area. Thus, in terms of mobility management, conventional schemes based on Radio Resource Management (RRM) measurements can be applied directly [10]. However, the challenges appear when dealing with MEO or in particular LEO satellites due to them being seen as moving base stations. Thus, the environment is more dynamic due to the moving satellites apart from the movement of the mobile user. Consequently, similarly to the cellular systems, the Earth stations/gateways maintain several registers for the satellite networks, such as [9]: *Home Location Register (HLR)* which stores the static subscriber data along with their current location, *Visitor Location Register (VLR)* which maintains the last known location of the mobile user, and *Satellite User Mapping Register (SUMR)* which stores the mapping of the mobile users to their current satellites along with the updated position of the satellites. A mobile user can register to the satellite network by sending an initial signal towards the available satellites that report this event to an Earth station/gateway. By using the information on the satellite's location from the SUMR, the Earth station/gateway can determine the mobile user's location. The HLR and VLR are interrogated about the user data and the SUMR is updated accordingly.

When a call is initiated towards a mobile user within the satellite network, it is first received by the Earth station/gateway which is able to localize the mobile user from the data in HLR and VLR while the information from the SUMR is used to locate the appropriate satellite to establish the communication.

Similarly to the mobility management in the cellular systems, also for satellite systems a hierarchical approach can be adopted where Registration Areas (RA) information is also maintained on the core network for the mobile user. The RA includes the Tracking Area (TA) information of the mobile user. Furthermore, the TA consist of either a set of moving satellite coverage beams/cells and is referred to as *mobile TA* or it can be defined by fixed areas on Earth referred to as *fixed TA* [10]. As expected, the mobile TA would introduce new challenges related to the moving beams that could force the mobile user to perform the registration process too often. Consequently, the fixed TA is the preferred one. The main purpose of the TA is to reduce the paging overhead when localizing the mobile user within the satellite networks.

Looking back at the typical satellite system example in Figure 5.5, and assuming two mobile users of the satellite network communicate by exchanging data, there are two possible scenarios for the routing of the data traffic, such that the transmitting mobile user will send the data to their corresponding satellite. Here, if the satellite is equipped with ISL then it will forward the data to the satellite responsible for the receiver mobile user via the satellite network. However, if the satellite is not equipped with ISL, then the data is forwarded to an Earth station/gateway and the routing takes place on the fixed terrestrial network until the Earth station/gateway responsible for the satellite above the receiver mobile user is reached. The data will then be forwarded to the corresponding satellite and sent back down to the receiver. Consequently, the integration of ISL enables a lower latency as the number of times the data travelling to Earth and back to the satellite is drastically reduced [9].

When the mobile user is actively connected to the satellite network, the mobility management also depends on the type of the satellite, bent pipe or on board processing. Consequently, for the bent pipe satellite, the mobility management can be further split into two categories, such as: (1) *user link handover* when the mobile user switches from one spot beam to another spot beam also referred to as *intra-satellite handover* as the mobile user is still within the footprint of the same satellite but a different cell; (2) *feeder link handover* when the satellite switches from one Earth station/gateway to another, also referred to as *gateway handover* as the mobile station is still in the footprint area of a satellite but the Earth station/gateway leaves the footprint. For the on board processing types of satellites they mainly include user link handover. The user link handover can also be an *inter-satellite* handover when the mobile user will handover from one satellite to another satellite while leaving the footprint of one satellite. Another type of handover would be the *inter-system handover* where the handover occurs from the satellite network to a terrestrial cellular network.

5.5 Practical Use-Case Scenario: Satellite Communications Using Altair WinProp

This section describes several practical use-case scenarios of the application of satellite communications using Altair WinProp.

5.5.1 Geostationary Communication Satellite for Rural/ Suburban Coverage

As we have previously seen, in rural rural/suburban areas where there is a low density of buildings the radio propagation mainly depends on the topography and the land usage or clutter. The aim of this practical use-case scenario is to use WinProp in order to investigate the coverage area offered by a geostationary satellite in a rural scenario [18]. To be able to achieve this, we are going to use a detailed topographical database and the land usage or clutter database (water, roads, buildings, forest, etc.) of the Stuttgart area as illustrated in Figure 5.10. It can be noted from the land usage that the central area of the map is represented by the city and the industrial zone, with residential areas spread across the map while a good portion of the map is represented by forested hills.

FIGURE 5.10
The Clutter/Morpho (Land Usage) Map for Rural/Suburban Scenario

We are going to use ProMan to study the coverage area provided by a geostationary communication satellite for this scenario. The overall workflow in ProMan is as follows:

- load a Rural/Suburban Scenario and consider additional land usage (clutter, morpho)

- load the three databases for the rural/suburban area: topography (pixel database), Land usage/Clutter (Pixel Database), Land usage/Clutter (Class definition).

- define type of transmitter, e.g., geostationary satellite.

- specify satellite characteristics.

- set up simulation, including selection of a radio propagation model.

- run simulation.

- inspect results.

In this practical use-case scenario one geostationary satellite was deployed at a height of 36000 km above sea level, operating at a carrier frequency of 2 GHz with a transmit power, EIRP of 90 dBm from an hypothetical isotropic antenna. The empirical two-ray radio propagation model was used with the added consideration of the knife edge diffraction to improve the accuracy of the model especially over the shadow areas. The predictions are computed at a height f 1.5 m above the ground level. Figure 5.11 illustrates the predicted received power and the LoS information. It can be noted that the received power and the LoS information follows the topography of the area. Consequently, the hill area where the forest is more dense and there is no LoS the received power is low, while the areas with direct LoS correspond to the open agriculture areas where also the received power is high.

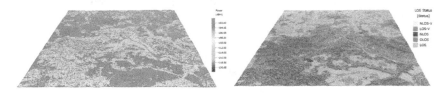

FIGURE 5.11
Received Power Prediction and Line of Sight

5.5.2 Geostationary Communication Satellite for Urban Coverage

The aim of this practical use-case scenario is to use WinProp in order to investigate the coverage area offered by a geostationary satellite in an urban scenario. To be able to achieve this, we are going to use first WallMan to generate a database map containing the geometry of a given urban area. Finally, the geometry from WallMan will be used by the main radio propagation simulation tool ProMan.

The workflow in WallMan is as follows:

- convert/import the file from another tool (e.g., OpenStreetMap[3])

- make optional modifications: building shapes, towers, courtyards, vegetation, etc.

- save for use in ProMan

For the purpose of this practical use-case we have used OpenStreetMap to obtain a database file to be converted in WallMan. The urban area we have selected is located in London central area, United Kingdom. However, it is possible to select any area you wish to investigate. The *.osm* file obtained from OpenStreetMap will be converted using WallMan (e.g., File > Convert Urban Database > Vector Database). Additional optional settings could be used like: check fill objects and check display building heights in status bar. Figure 5.12 illustrates the outcome of this step which can be saved as an *.odb* file. Normaly, this step would have been enough and the *.odb* database file would have been ready to be used in ProMan.

---

[3] OpenStreetMap: www.openstreetmap.org

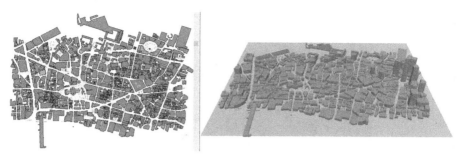

FIGURE 5.12
WallMan Urban Database – London Central Area Map

However, in this use case we are going to use Intelligent Ray Tracing (IRT) as radio propagation model. Consequently, we need to pre-process the *.odb* file before using it in ProMan. For this a new project is created (File > New Project) that will use the *.odb* file we just created. We then use the Preprocessing > Edit Preprocessing Parameters option to select the mode for the pre-processing run for the IRT propagation model. The output would be the *.oib* database file representing the Outdoor IRT Binary database. This file is generated by selecting *Preprocessing > ComputeCurrentProject* in the menu and it contains building data as well as the visibility information along with the considered parameters for this specific pre-processing. This file is now ready to be used in ProMan with the corresponding propagation model, in this case IRT.

The geometry representing the London central area was generated in Wall-Man with the pre-processing database file, *.oib* corresponding to the IRT propagation model. We will now use this database files in the propagation simulation tool ProMan to study the coverage area provided by a geostationary communication satellite within the urban environment.

The overall workflow in ProMan is as follow:

- load an Urban Scenario

- load the preprocessed vector database that was prepared in WallMan.

- define type of transmitter, e.g., geostationary satellite.

- specify satellite characteristics.

- set up simulation, including selection of a radio propagation model.

- run simulation.

- inspect results.

Similarly to the previous use-case scenario, in this practical use-case scenario one geostationary satellite was deployed at a height of 36000 km above sea level, operating at a carrier frequency of 2 GHz with a transmit power,

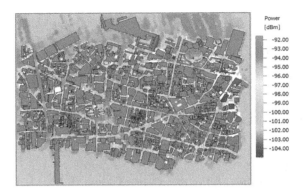

FIGURE 5.13
Received Power Prediction – London Central Area Map

EIRP of 90 dBm from an hypothetical isotropic antenna. The ICT radio propagation model was used and the predictions are computed at a height of 1.5m above the ground level. Figure 5.13 illustrates the predicted received power within the urban scenario. It can be noted that the buildings cast shadows northward, which indicates that the transmitter is located south of the city. The white areas indicate that there is no signal reception in those parts.

5.5.3 GPS Satellites System for Urban Coverage

The aim of this practical use-case scenario is to use WinProp in order to investigate the coverage area offered by a GPS satellite system in an urban scenario. To be able to achieve this, we are going to use the same urban database of London central area as per the previous use-case scenario and as indicated in Figure 5.12. We need to provide the GPS orbit data information, and this is possible through the use of two methods, such as: (a) Almanac datasets or (b) TLE datasets both downloaded from CelesTrak[4]

In this practical use-case scenario we are going to use the Almanac dataset that consists of a subset of reduced precision GPS satellite clock and ephemeris data in YUMA format, an easy to read format of the Almanac data. Figure 5.14 illustrates one record sample from the YUMA Almanac dataset used in this use-case scenario. There are 32 records in this file raging from PRN 1 to 32 with the parameters identified within the message structure. Consequently, the entries for ID, Health (e.g.,mapped to three health categories, such as: active, bad and dead), and Week are represented in decimal format.

The geometry representing the London central area was generated in WallMan with the pre-processing database file, *.oib* corresponding to the IRT propagation model. We will now use this database files in the propagation

---
[4]CelesTrak – http://www.celestrak.com/GPS/

```
******** Week 159 almanac for PRN-01 ********
ID:                    01
Health:               000
Eccentricity:          0.1182317734E-001
Time of Applicability(s): 589824.0000
Orbital Inclination(rad): 0.9879159848
Rate of Right Ascen(r/s): -0.7428880871E-008
SQRT(A) (m 1/2):          5153.629883
Right Ascen at Week(rad): 0.3117665025E+001
Argument of Perigee(rad): 0.898018793
Mean Anom(rad):           0.1470184639E+001
Af0(s):               0.3728866577E-003
Af1(s/s):             -0.7275957614E-011
week:                 159
```

FIGURE 5.14
YUMA Almanac Data Sample

simulation tool ProMan to study the coverage area provided by a GPS satellite system defined by the Almanac dataset, within the urban environment.

The overall workflow in ProMan is as follow:

- load an Urban Scenario

- load the preprocessed vector database that was prepared in WallMan.

- define type of transmitter, e.g.,Satellite > GPS Satellites (Almanach Definitions).

- specify the name and location of the Almanach file.

- set up simulation, including selection of a radio propagation model.

- run simulation.

- inspect results.

When processing the Almanach file, ProMan enables you to set the time instance (GMT) and a minimum elevation angle threshold so that the satellites below this elevation threshold will not be converted. Consequently, the elevation angle threshold was set to $5°$. With these setups ProMan converted 11 transmitter satellite antennas operating on the frequency of 1575.42 MHz with a transmitter power of 26.607 Watts. The ICT radio propagation model was used and the predictions are computed at a height of 1.5m above the ground level. Figure 5.15 illustrates the predicted received power of one of the satellite transmitters that is not directly over the urban scenario. It can be noted that the buildings cast shadows southward, which indicates that this particular transmitter is located north of the city. The white areas indicate that there is no signal reception in those parts.

FIGURE 5.15
Received Power Prediction GPS Satellite – London Central Area 2D and 3D Maps

Bibliography

[1] William Stallings. *Wireless Communications & Networks: Pearson New International Edition PDF eBook*. Pearson Higher Ed, 2013.

[2] Husam Al-Deen F Kokez. On terrestrial and satellite communications for telecommunication future. In *2020 2nd Annual International Conference on Information and Sciences (AiCIS)*, pages 58–67. IEEE, 2020.

[3] Shanzhi Chen, Ying-Chang Liang, Shaohui Sun, Shaoli Kang, Wenchi Cheng, and Mugen Peng. Vision, requirements, and technology trend of 6g: How to tackle the challenges of system coverage, capacity, user data-rate and movement speed. *IEEE Wireless Communications*, 27(2):218–228, 2020.

[4] Mehdi Rasti, Shiva Kazemi Taskou, Hina Tabassum, and Ekram Hossain. Evolution toward 6g wireless networks: A resource management perspective. *arXiv preprint arXiv:2108.06527*, 2021.

[5] Amin Shahraki, Mahmoud Abbasi, Md Piran, Amir Taherkordi, et al. A comprehensive survey on 6g networks: Applications, core services, enabling technologies, and future challenges. *arXiv preprint arXiv:2101.12475*, 2021.

[6] Aygün Baltaci, Ergin Dinc, Mustafa Ozger, Abdulrahman Alabbasi, Cicek Cavdar, and Dominic Schupke. A survey of wireless networks for future aerial communications (facom). *IEEE Communications Surveys & Tutorials*, 2021.

[7] Mustafa Ergen, Feride Inan, Onur Ergen, Ibraheem Shayea, Mehmet Fatih Tuysuz, Azizul Azizan, Nazim Kemal Ure, and Maziar Nekovee. Edge on wheels with omnibus networking for 6g technology. *IEEE Access*, 8:215928–215942, 2020.

[8] Behrouz A. Forouzan. *Data Communications and Networking.* McGraw-Hill, Inc., USA, 3 edition, 2003.

[9] Jochen H Schiller. *Mobile communications.* Pearson education, 2003.

[10] Shanzhi Chen, Shaohui Sun, and Shaoli Kang. System integration of terrestrial mobile communication and satellite communication—the trends, challenges and key technologies in b5g and 6g. *China Communications,* 17(12):156–171, 2020.

[11] Oltjon Kodheli, Eva Lagunas, Nicola Maturo, Shree Krishna Sharma, Bhavani Shankar, Jesus Fabian Mendoza Montoya, Juan Carlos Merlano Duncan, Danilo Spano, Symeon Chatzinotas, Steven Kisseleff, et al. Satellite communications in the new space era: A survey and future challenges. *IEEE Communications Surveys & Tutorials,* 23(1):70–109, 2020.

[12] John V Evans. Satellite systems for personal communications. *Proceedings of the IEEE,* 86(7):1325–1341, 1998.

[13] Paolo Chini, Giovanni Giambene, and Sastri Kota. A survey on mobile satellite systems. *International Journal of Satellite Communications and Networking,* 28(1):29–57, 2010.

[14] Jonathan C McDowell. The low earth orbit satellite population and impacts of the spacex starlink constellation. *The Astrophysical Journal Letters,* 892(2):L36, 2020.

[15] J Tregloan-Reed, A Otarola, E Ortiz, V Molina, J Anais, R González, JP Colque, and E Unda-Sanzana. First observations and magnitude measurement of starlink's darksat. *Astronomy & Astrophysics,* 637:L1, 2020.

[16] Ahmed El-Rabbany. *Introduction to GPS: the global positioning system.* Artech house, 2002.

[17] Cory Beard and William Stallings. *Wireless communication networks and systems.* Pearson, 2015.

[18] Altair. Altair winprop 2021.1 user guide. *Altair Engineering Inc.,* 2021.

Chapter 6

Wireless Evolution

"My designs were so deceptively simple that it was easy for people to assume I just had easy problems, whereas others, who made super-complicated designs (that were technically unsound and never worked) and were able to talk about them in ways that nobody understood, were considered geniuses."

Radia Perlman
Computer scientist and Network Engineer for Dell EMC,
Developed the algorithm behind the Spanning Tree Protocol.

6.1 Wireless Technologies Evolution

Wi-Fi (Wireless Fidelity) or the Institute for Electrical and Electronics Engineers (IEEE) standard 802.11 [1], was introduced for the first time in 1997. The design goals for Wireless Local Area Network (WLAN) were to enable global and seamless operation with low power for prolonged battery use, without any special permissions or licenses needed to deploy and use the wireless network, enable a robust transmission technology, with simple management and easy to use by everyone, with advanced security, privacy and safety as well as to enable transparency concerning applications and higher layer protocols. The standard was initially able to provide 1 or 2 Mbps bit rate, using the 2.4 GHz frequency band. Over the years the technology evolved and there were several amendments made to the original standard and a set of standards for wireless area networks specified by IEEE emerged. However, the standard contains information about the physical and medium access layers only and it operates in the license free ISM band at 2.4 GHz and at 5 GHz. However, the IEEE 802.11 family consists of various standards with different features and capabilities.

Table 6.1 presents a list of the most important (released or in progress of being released), amendments and supplements of the IEEE 802.11 family.

The first improved version of the original standard was **IEEE 802.11a** [2] which could offer an increase in throughput up to 54 Mbps using the 5 GHz frequency band. Because of the low number of devices operating at 5 GHz, at that time, the new amendment provided less interference. However, the biggest disadvantage was the short coverage area and the quick degradation of the signal due to strong shading at higher frequency and no QoS provisioning. IEEE 802.11a supports multiple data rates of 6, 9, 12, 18, 24, 36, 48, and 54 Mbps depending on the SNR, using Orthogonal Frequency Division Multiplexing (OFDM) modulation technique. Even though the standard provides a theoretical data rate of up to 54 Mbps, in a real scenario, the approximate received throughput is 25 Mbps [3].

Another well-known amendment and probably the most deployed was **IEEE 802.11b** [4], operating at 2.4 GHz frequency and providing data rates up to 11 Mbps using Complementary Code Keying (CCK) modulation. The standard provides four theoretical data rates of 1, 2, 5.5, and 11 Mbps, depending on SNR. In practice, for IEEE 802.11b the approximate received throughput is 6 Mbps [3]. The major advantage of IEEE 802.11b is its popularity being integrated in most of the devices with many installed systems available worldwide. However, the disadvantage is the use of the free ISM band with heavy interference, slow speed and also no QoS provisioning.

Similarly to IEEE 802.11b, the next ratified amendment, **IEEE 802.11g** [5], also gained widespread adoption because of the increase in throughput, which could go up to 54 Mbps while operating at 2.4 GHz frequency. It provides backwards compatibility with IEEE 802.11b devices. The standard uses OFDM and CCK modulation techniques and provides data rates of 1, 2, 5.5, 6, 9, 11, 12, 18, 24, 36, 48, and 54 Mbps depending on the SNR. The approximate received throughput in a real scenario is 22 Mbps when there are only IEEE 802.11g clients in the network [3]. The user throughput is significantly lower in the presence of IEEE 802.11b devices. Similarly to IEEE 802.11b, the major disadvantage is the use of the crowded ISM band and the lower speed due to backward compatibility.

The next amendment was **IEEE 802.11n** [6] or Wireless Fidelity (Wi-Fi) 4, which brings improved reliability, more predictable coverage, improved immunity to noise, compatibility with IEEE 802.11a/b/g, and a higher throughput which can achieve performance parity with 100 Mbps fast Ethernet. The new amendment doubles the channel bandwidth to 40 MHz and works with MIMO leading to an increased data rate. The IEEE 802.11n operates in dual band at 2.4 GHz and 5 GHz using OFDM and CCK modulation techniques and providing data rates from 6.5 to 600 Mbps.

The next evolution led to Gigabit Wi-Fi with the **IEEE 802.11ac** amendment or Wi-Fi 5 which can achieve data rates of more than 7 Gbps. IEEE 802.11ac operates in the 5 GHz band only making use of up to 8x8 MIMO, increase in the channel bandwidth of up to 160 MHz and increase in the constellation order of Quadrature Amplitude Modulation (QAM) from 64 QAM to 256 QAM. The standard also uses Multi User-MIMO (MU-MIMO) with simultaneous beams to multiple stations and advanced channel measurements.

It also makes use of special RTS/CTS to check for legacy devices within the network [7].

Wi-Fi 6 is represented by **IEEE 802.11ax** amendment that makes a more efficient use of the spectrum to increase the user throughput up to 9.6 Gbps. This is done by adopting the OFDMA approach similarly to the cellular systems [8]. The standard operates in both 2.4 GHz and 5 GHz bands and enables MU-MIMO, as well as 1024-QAM.

The latest evolution to Wi-Fi 7 is represented by **IEEE 802.11be** that aims to improve the user experience in order to support more advanced applications like 8K video, VR, AR, real time gaming, remote office and cloud computing. This is done by doubling the bandwidth up to 320 MHz and the number of spatial streams in MU-MIMO (e.g., 16x16 MU-MIMO), enabling the use of 4K-QAM and introducing an enhanced OFDMA. Wi-Fi 7 operates in the 1-7.25 GHz bands including 2.4, 5 and 6 GHz and the maximum nominal throughput goes up to 46Gbps [7].

TABLE 6.1

Overview of IEEE 802.11 Supplements and Amendments

Standard	Description
802.11	• Data rates: 1 Mbps and 2 Mbps • Frequency: 2.4 GHz • Modulation: FHSS, DSSS and IR-Phy
802.11a	• Data rates: up to 54 Mbps • Frequency: 5 GHz • Modulation: OFDM
802.11b	• Data rates: up to 11 Mbps • Frequency: 2.4 GHz • Modulation: extension of DSSS
802.11c	Ensures wireless bridging between APs
802.11d	Specification for operation in additional regulatory domains
802.11e	Provides QoS Enhancements and prioritization of data packets
802.11F	Inter-Access Point Protocol, provides interoperability between multi-vendor APs

Continued on next page

TABLE 6.1

Overview of IEEE 802.11 Supplements and Amendments (Continued)

Standard	Description
802.11g	• Data rates: up to 54 Mbps • Frequency: 2.4 GHz compatible with 802.11b • Modulation: OFDM
802.11h	Radio Resource Management
802.11i	Enhanced Security
802.11j	Designed only for Japanese market, operating in 4.9 to 5 GHz
802.11k	Radio Resource Measurements.
802.11m	Performs maintenance, technical and editorial corrections and improvements.
802.11n	Provides higher throughput improvements by using MIMO technology. Data rates between 108 Mbps – 320 Mbps. (Wi-Fi 4)
802.11p	Support for Vehicular Environment
802.11r	Permits continuous connectivity by providing fast BSS transition.
802.11s	Support for mesh networking.
802.11t	Provides test methods and metrics.
802.11u	Interworking with non-802 networks such as cellular networks.
802.11v	Network management. Extensions of current management functions, channel measurements. Definition of a unified interface.
802.11w	Securing of network control. Classical standards like 802.11, but also 802.11i protect only data frames, not the control frames. Thus, this standard should extend 802.11i in a way that, e.g., no control frames can be forged.
802.11y	Extensions for the 3650–3700 MHz band in the USA
802.11z	Extension to direct link setup.
802.11aa	Robust audio/video stream transport.
802.11ac	Very High Throughput <6 GHz (Wi-Fi 5)
802.11ax	(Wi-Fi 6) High Efficiency Wireless: dense deployments (sports stadium, airports, etc.), throughput 4 x higher than 802.11ac and 802.11n.
802.11ad	Very High Throughput in 60 GHz.

Continued on next page

TABLE 6.1

Overview of IEEE 802.11 Supplements and Amendments (Continued)

Standard	Description
802.11ay	Next Generation 60 GHz, max throughput of at least 20 Gbps, improvement on 802.11ad (up to 7 Gbps), increase in range and reliability.
802.11ae	Prioritization of Management Frames.
802.11af	TV white space, ah: sub 1 GHz, ai: fast initial link set-up; aq: pre-association discovery.
802.11ah	Wi-Fi HaLow – uses 900 MHz license exempt bands to provide extended range Wi-Fi networks.
802.11az	Next Generation Positioning (NGP) – determination of absolute and relative position with better accuracy
802.11ba	"Wake-Up Radio" (WUR) – extending the battery life of devices and sensors within an Internet of Things network
802.11be	(Wi-Fi 7) introduces, forward-compatible physical layer (PHY), scalable sounding, Multiple Access Point (Multi-AP) cooperation.
802.11bd	enhancements of IEEE 802.11p in terms of throughput, latency, reliability and communication range.

The IEEE 802.11 Task Group "k" developed an important extension of the IEEE 802.11 wireless LAN standard, which is referred to as 802.11k [9] (ratified in 2008). This extension is defined for the provisioning of the radio resource measurement, in order to allow radio stations to request and exchange information about the usage of the wireless medium. The IEEE 802.11k standard defines basic structures for requesting and reporting measurements information, but only for IEEE 802 networks [10]. There are no interoperability methods between heterogeneous networks defined in IEEE 802.11k, and no inter-RAT measurements procedures. The IEEE 802.11k standard defines different types of measurements [11], including: the beacon report which provides information on signal strength and signal to noise ratio; the frame report, with information on all received frames; the channel load report that returns the channel utilization measurement (as observed by a measuring station); noise histogram report that provides the expected value of noise collected in a specific number of channels in the measurement duration; statistic report with information related to link quality and network performance; location report that contains the current location formatted based on the IETF RFC 3825 standard [12], in terms of latitude, longitude and altitude; neighbor report that provides information about the neighbors of the associated AP; and link measurement report that provides the instantaneous quality of a link. However, the IEEE 802.11k does not include any radio resource management; the objective is to provide radio resource measurements. The standard contains two main message types: request and report messages. Radio stations can exchange messages in two ways: station-to-station or station-to-AP. These messages can be sent in unicast, broadcast or multicast nodes. Each request/report message is included in an action frame with the category field set to radio measurement, and has information about the requested measurement settings (channels, duration, start time, etc). The frame contains the measurement information (in request) or the measurement results (in reports). The action frame is then carried by a MAC management frame. The standard does not specify the default measurement duration and allows each radio station to specify duration along with a measurement request. The requested radio station can decide on the measurement duration and also whether or not to repeat the measurements after a certain time. The IEEE 802.11k standard allows the inclusion of multiple measurement elements in one measurement request or report. The standard provides information about the current location but the acquisition mechanism for positioning itself is not included in the standard [13].

It is worth mentioning that not all the standards will end up being implemented in products, and many ideas remain as discussions at a working group level.

The IEEE 802.11 standard can operate in two modes: (1) **infrastructure network mode** or non-ad-hoc networks – meaning that the communication between mobile clients is done through a central component (e.g., Access Point (AP)). This type of network offers the advantage of scalability and centralized management and (2) **ad-hoc network mode** – where the communication between mobile clients is done through other mobile clients used by the routing mechanism for data forwarding. This type of network is decentralized and it does not rely on pre-existing infrastructure. In ad-hoc network mode the communication between devices is happening directly and they form a so called Independent Basic Service Set (IBSS). Its advantage is that it eliminates the cost of adding a central component.

The basic architecture of an infrastructure IEEE 802.11-based network is illustrated in Figure 6.1. The typical IEEE 802.11 network consists of one or more stations (STAs) and one AP, referred to as the Basic Service Set (BSS). The STAs are represented by any user equipment with a WLAN interface, including: laptops, Personal Digital Assistant (PDAs), smartphones, etc. Nowadays most of the smartphones have both cellular and WLAN interfaces included. The APs can be connected to the same distributed system (DS) or wired network and this configuration is referred to as the Extended Service Set (ESS). The IEEE 802.11 standard allows the stations to roam within the ESS.

Figure 6.2 presents a possible WLAN configuration where a wireless device is connected to an AP which is connected to the infrastructure network or LAN where a wired device resides. The protocol stack of the entities involved

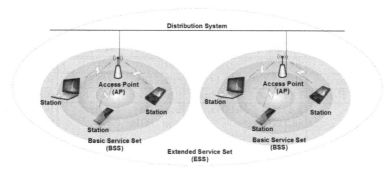

FIGURE 6.1
WLAN Basic Architecture

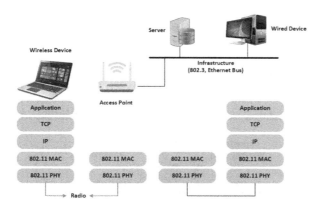

FIGURE 6.2
WLAN Basic Configuration Setup

are illustrated. It can be noted that the standard covers the physical layer (PHY) and medium access layer (MAC) and only the end devices require the full stack implementation, with the higher layers like application, TCP, IP. We have seen that depending of the wireless standard used the PHY can have different characteristics. A summary of the PHY characteristics of the wireless generations is presented in Table 6.2.

The MAC layer handles the access mechanisms, the user data fragmentation as well as the encryption of data and consists of two MAC sublayers, such as: (1) Distributed Coordination Function (DCF) – its implementation is mandatory and is used in IEEE 802.11 infrastructure and ad-hoc configuration modes with bursty traffic. It makes use of Carrier Sense Multiple Access with Collision Avoidance (CSMA/CA) as the access method and (2)

TABLE 6.2
Wireless Generations Characteristics

	Wi-Fi 4	Wi-Fi 5	Wi-Fi 6	Wi-Fi 6E	Wi-Fi 7
Launch Date	2007	2013	2019	2021	2024
IEEE standard	802.11n	802.11ac	802.11ax	802.11ax	802.11be
Max. Data Rate	1.2Gbps	3.5 Gbps	9.6 Gbps	9.6 Gbps	46 Gbps
Bands	2.4 GHz and 5 GHz	5 GHz	2.4 GHz and 5 GHz	6 GHz	1-7.25 GHz (including 2.4 GHz, 5 GHz and 6 GHz bands)
Security	WPA 2	WPA 2	WPA 3	WPA 3	WPA 3
Channel size	20, 40 MHz	20, 40, 80, 80 + 80, 160 MHz	20, 40, 80, 80 + 80, 160 MHz	20, 40, 80, 80 + 80, 160 MHz	up to 320 MHz
Modulation	64-QAM OFDM	256-QAM OFDM	1024-QAM OFDMA	1024-QAM OFDMA	4096-QAM OFDMA (with extensions)
MIMO	4x4 MIMO	4x4 MIMO, DL MU-MIMO	8x8 UL/DL MU-MIMO	8x8 UL/DL MU-MIMO	16x16 MU-MIMO

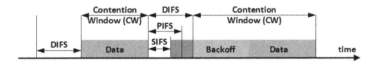

FIGURE 6.3
Elements of CSMA

Point Coordination Function (PCF) – its implementation is optional and is used for transmitting high priority and time sensitive data in IEEE 802.11 infrastructure mode.

The access method implemented in wired networks is Carrier Sense Multiple Access with Collision Detection (CSMA/CD) which cannot be used efficiently in wireless networks, because a station might not be able to detect collision due to the hidden terminal problem or the increased distance between the receiver and transmitter that might prevent the station form hearing the collisions at the other end due to the fading of the signal. Thus, in wireless networks CSMA/CA is used instead. The elements of CSMA in IEEE 802.11 are illustrated in Figure 6.3 and consist of three parts: (1) the Inter-Frame Spacing (IFS) that further consists of three types of frames that differ in duration. The shorter the IFS the higher the transmission priority. Thus, the three

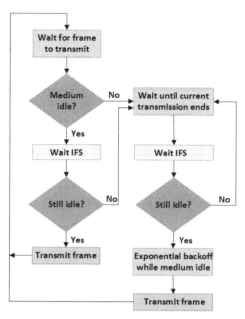

FIGURE 6.4
DCF Algorithm

IFS type of frames are: (a) *Short IFS (SIFS)* – used to transmit high priority signalling messages like ACK, CTS, etc.; (b) *PCF IFS (PIFS)* – used by the AP in the PCF mode when issuing polls; (c) *DCF IFS (DIFS)* – used as the minimum delay for asynchronous frames contending for access; (2) Contention Window (CW) which is used for contention and transmission and (3) Back-off Mechanism which is part of the CW and is used when multiple stations compete for access. The mechanism will generate a random number referred to as back-off that indicates how long the next transmission is deferred.

Figure 6.4 [14] indicates the flow chart of the DCF algorithm. Whenever a station has a frame to transmit it will first sense to medium to see if it is *idle* or *busy*. If the medium is *idle* the station will wait for the duration of an IFS and checks again. If the medium is still *idle* the station will transmit immediately. If the medium is observed as being *busy*, the station will defer the transmission and will continue to monitor the medium until the current transmission ends after which it waits another IFS. After the IFS has ended, the station will check again to see if the medium is still *idle*, in which case the station will back-off a random amount of time before transmitting [15].

An example of DCF in a 802.11 network is illustrated in Figure 6.5. In this example, there are five stations (e.g., Sender A, B, C, D and E) competing for access to the wireless medium [16]. Sender A is already sending while Senders B, C and D sense the channel and find it busy, thus they will get to back-off a random amount of time before sensing the channel again. After sender

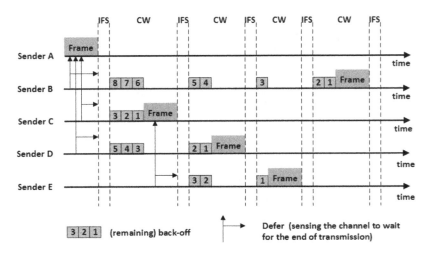

FIGURE 6.5
DCF Example in 802.11

A has finished the transmission, senders B, C, and D will wait for IFS and start their counters. Sender's C back-off time finished first and can start the transmission after making sure the medium is idle while senders B and D will freeze their counters. A fifth sender, E wants to transmit as well and senses the channel busy. Thus, sender E will calculate the back-off timer as larger than the reminder of D and smaller than the reminder of B. After sender C has finished the transmission, senders B, D and now E will wait for IFS and start their counters and the process repeats.

The DCF algorithm can be extended to integrate the short control frames of RTS and CTS in order to avoid the *hidden terminal problem*. In this situation when a sender has data to transmit it will wait for DIFS before issuing an RTS as the RTS packets are not given any higher priority as compared to other regular packets. The RTS packet will include information about the intended receiver of the data and the estimated duration of the entire transmission so that all the other stations that receive the RTS to update their NAV indicating the earliest point in time at which the station can try to access the medium again. Upon the receipt of the RTS, the receiver will wait for SIFS before issuing a CTS, meaning that CTS is given higher priority. The CTS is received by all the stations in the range of the receiver and includes information on the intended communication pair and the estimated duration of the transmission. The sender will wait for SIFS before start sending the data indicating higher priority while the receiver will wait for another SIFS before sending the ACK indicating the correct reception of the data.

When dealing with time sensitive transmissions, DCF fails to guarantee a maximum access delay or a minimum transmission data rate. Consequently, the PCF mode can be used in these situations, on top of the DCF. An example

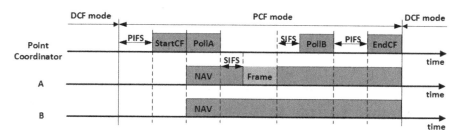

FIGURE 6.6
PCF Polling Example in 802.11

of PCF polling is illustrated in Figure 6.6. In this situation the AP will act as the point coordinator that controls the medium access and polls every station. Thus, when transitioning from DCF to PCF the AP will sense the medium and if idle will wait for PIFS before starting the PCF mode and start polling the stations subsequently for data transmission. If a station has no data to transmit, the AP will wait for SIFS or PIFS before polling the next station or terminating the PCF mode, respectively.

The stations that belong to an infrastructure mode 802.11 WLAN can manage their power through their *Power Save Mode (PSM)* which is based on synchronous sleep scheduling policy. In order to conserve energy, the wireless stations can alternate between the active mode and the sleep mode. Consequently, when a wireless station wants to enter the PSM mode, it has to inform the AP that it has the PSM enabled by using the power management bit within the frame control field of the transmitted frames. Once in PSM, the wireless station switches the NIC card to a lower power sleep state and wakes up periodically to receive beacons from the AP. All the packets intended for the wireless station are buffered by the AP which will indicate if a pending packet exists using the Traffic Indication Map (TIM). These TIMs are included within the beacons that are periodically transmitted from the AP. On receipt of an indication of a waiting packet at the AP, the wireless station sends a power save (PS) poll frame to the AP and waits for a response in the active state. The AP responds to the poll by transmitting the pending packet or indication for future transmission. In the case of multicast and broadcast packets, the AP transmits an indication for pending packets using a delivery TIM (DTIM) beacon frame, immediately followed by the actual multicast or broadcast packet, without any explicit PS poll from the stations. On the other side, whenever a PSM station has data to transmit, it simply wakes up, sends its packet, and then resumes its sleep schedule protocol as appropriate.

The handover in 802.11 is a mobile controlled handover and we can identify three types of handover, such as: (1) *no handover* when the terminal is static or it moves within the same BSS; (2) *BSS handover* when the terminal moves from one BSS to another BSS within the same ESS and (3) *ESS handover* when a terminal moved from one BSS to another BSS that is part of a new

ESS. In this situation the handover must be supported by Mobile IP to avoid disconnections.

6.2 Mobile Ad-Hoc Networks

The mobile ad-hoc networks (MANETs), are represented by networks of wireless devices that do not necessarily rely on an existing infrastructure and the communication is done directly between the terminals when required. We have seen until now that the mobility support relies on the existence of the infrastructure, for example the BSs in the cellular systems, or the APs in the wireless network. However, there are some situations where the infrastructure is not available or it is too expensive to deploy. Consequently, for these scenarios MANETs are the best option. MANETs are mobile and make use of the wireless communication. Because the communication can happen through single-hop where all the stations are at maximum one hop distance apart or through multi-hop that can offer a larger coverage area, and all these hops are mobile nodes, MANET represents an extended concept of mobility, which is the network mobility similar to having a network of moving routers [16]. Consequently, one of the biggest challenges in MANETs is the routing of data as there is no default route available and the routing replies on the fact that every station should be able to forward so that single or multi hop communication is possible.

An example of routing in MANET is illustrated in Figure 6.7 [16]. All the stations in MANET are mobile and are also able to forward data, which makes MANET's topology dynamic and rapidly changing from one time frame to another. Moreover, different stations have different capabilities which makes the communication links between the stations asymmetric. This means that if a station has a good communication link for downlink, that is not necessarily the case for the uplink communication link as well, which might be weak or nonexistent. This can be due to different transmit power capabilities of the transmitting station, or that the topology changed and the stations are closer of further apart to be able to communicate anymore due to broken routing links. However, when more stations are in very close proximity, there could be too many redundant links, and they could also suffer from interference.

Traditional routing algorithms are built for wired networks, which makes them unsuitable for MANETs, as seen in Figure 6.7. For example, the *Distance Vector Routing* protocol that is often used in wired networks works based on the idea that each stations will store a routing table that has an entry to each destination (e.g., destination, distance, neighbour station) and each station will maintain the distance information to very other node. In case a station receives an update message from a neighbour with changes in the neighbourhood, it will update its routing table accordingly and will also send

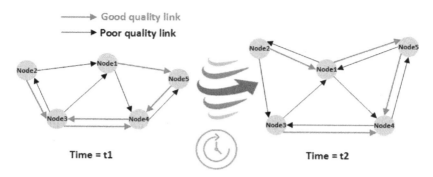

FIGURE 6.7
MANET Routing Example

an update to all of its neighbours. The advantages of this routing protocol is that the messages follow the shortest path and that it only sends updates when there is a change in the topology. However, some of the topology changes might be irrelevant for a given communication pair with stations still being required to update their routing table entries. Consequently, every station might end up storing a big routing table which consumes memory, processing time, power, and generates additional overhead traffic on the network.

Another traditional routing algorithm often used for wired networks is the *Link-State Routing* protocol, which works based on the idea that all stations in the network need to get a complete picture of the network through periodic notifications of all stations about the current network state of all the physical links. The only advantage of this protocol is that the messages follow the shortest path. However, the main disadvantages are that every station will have to store the whole graph of the network even fro the links that are not on any path. Moreover, every station needs to send and receive regular messages about the current status of the whole graph which adds significant overhead traffic into the network.

Consequently, the traditional routing algorithms that have been designed for fixed networks where there are infrequent changes in topology and symmetric links cannot be used for MANET where the topology is dynamic with frequent changes in the connections, the connection quality and also the stations involved in the communication. For these reasons there are many routing protocols proposed in the current literature, especially for MANETs that aim to overcome the challenges associated with the distance vector routing protocol and the link state routing protocol. In general, the existing routing protocols for MANETs are classified into three main categories, such as [17; 17]: *proactive protocols* which maintain the routes between every communication pair at all times by using periodic updates; *reactive protocols* which determine the routes only if and when needed by using a route discovery process initiated by the source; *hybrid protocols* which combine the basic properties of proactive and reactive protocols making them more adaptive.

6.2.1 Proactive Routing Protocols

The proactive routing protocols rely on stations maintaining up to date routing tables of known destinations. The advantage of these protocols is that they enable lower latency since the routes are maintained at all times in the routing tables. However, this comes at the cost of continuous update messages to neighbours which increases the network traffic overhead. This makes them unsuitable for highly dynamic networks with frequent topology changes, as the routing tables will require to be updated with every topology change leading to degrading network performance especially at high loads due to the additional traffic overhead generated.

One example of proactive routing protocol is the *Destination-Sequenced Distance Vector (DSDV)* algorithm [18] that represents an extension of the distance vector routing protocol which avoids loops and inconsistencies by introducing sequence numbers for all routing updates and also ensures in-order execution of all updates. The temporary routing loops that the distance vector routing protocol is prone to are formed when an error occurs in a group of stations and the path to a certain destination forms a loop. This is overcome by DSDV by having each station in the network maintain a routing table storing: destinations, next hop addresses, number of hops, as well as the sequence numbers. Moreover, DSDV reduced the amount of overhead in the network by limiting the periodical routing table updates to the size of one packet containing only new information [19]. However, in case of a major topology change DSDV can add significant control traffic overhead into the network making it unsuitable for large networks.

The *Wireless Routing Protocol (WRP)* [20] is another example of proactive routing protocols that makes use of the predecessor information to avoid temporary routing loops. However, the stations suffer from a significant memory overhead as they require to maintain four routing tables. Hello messages are transmitted to maintain connectivity between neighbouring stations where there was no recent packet transmission recorded. Thus, the stations cannot go into sleep mode and are required to stay awake at all times increasing on one side their energy consumption and the network traffic overhead on the other side.

6.2.2 Reactive Routing Protocols

In contrast to the proactive routing protocols, the reactive routing protocols use a route discovery process that determines the routes only if and when needed lowering the network traffic overhead. However, because a route from a station X to a station Y will be found only when station X attempts to send to station Y, this will cause higher latency in data transmission. Moreover, as compared to the proactive routing protocols, the reactive protocols are more suitable within highly dynamic environments. However, which approach achieves a better trade-off between the additional overhead vs. the latency of route discovery, depends on the traffic and mobility patterns.

An example of reactive routing protocol is Dynamic Source Routing (DSR) [21] that consists of three phases: (1) the route discovery phase - which is initiated by the source station only when there is data to be sent towards a destination station. The route discovery process consists of flooding the network with route requests messages that are forwarded by the stations after appending their own identifier to the message. (2) the route reply phase - once the intended destination station receives the route request message it will answer by sending a route reply message on the route obtained by reversing the route appended to the received route request message. (3) the data forwarding phase - upon the receipt of route reply message the source station will begin data transmission towards the destination station on the identified route. DSR makes use of route caching such that during the route discovery process, the stations involved in the process will learn and store in their route cache multiple routes. This can speed up the route discovery process and can also reduce the propagation of route requests by having stations using their local route cache. Another advantage of DSR is that it reduces the overhead of route maintenance by maintaining the routes only between the stations that need to communicate. However, as stations append their own identifier to the route request message the packet header size grows with the route length adding overhead into the network. Moreover, the flooding of route requests might reach all the stations within the network. DSR can also cause a *route reply storm* problem when too many route replies come back from stations using their local route cache. Another issue might arise when stations use their stale cached route that could eventually pollute other caches.

Ad Hoc On-Demand Distance Vector (AODV) [22] is another reactive routing protocol that tries to overcome the shortcomings of DSR. We have seen that DSR can lead to large packet headers as stations append their own identifiers to it which actually degrade the performance of the system especially if the payload data of a packet is small. Consequently, AODV introduces temporary routing tables at the stations for the duration of the communication reducing the routing overhead. However, AODV also consists of the three phases as defined by DSR, with the difference that the route replies only carries the destination IP address and the sequence number to enable loop free routing similar to DSDV. Similarly to DSR, AODV will only maintain the routes between the communicating stations and also assumes symmetric bidirectional links.

6.2.3 Hybrid Routing Protocols

The hybrid routing protocol represents a compromise between the proactive and reactive routing protocols. Thus, the hybrid approach will try to reduce the additional overhead introduce by the proactive routing while also reducing the latency of route discovery of the reactive routing. This is achieved by maintaining some form of routing table for near by nodes and using the route discovery to determine the routes for far away nodes.

An example of a hybrid routing protocol is the Zone Routing Protocol (ZRP) [23] that makes use of routing zones defined as a range in hops. The stations within these routing zones will maintain the network connectivity in a proactive manner, meaning that all routes are available. While in the case of the stations outside the routing zones the routes are determined in a reactive manner if and when required. In this way, ZRP reduces the amount of control traffic overhead especially under large networks when compared to the proactive protocols. Additionally, the delay is also reduced when compared to the reactive protocols as ZRP can discover the routes faster. However, the challenge of ZRP is the size of the routing zone, as large values can lead to similar performance of proactive protocols while small values lead to the performance of a reactive protocol. Thus, a trade-off needs to be found.

6.3 Vehicular Networks

A special class of MANETs are Vehicular Ad-Hoc Network (VANET) or vehicular networks, where the mobile stations are in fact vehicles moving at a fast speed under a highly dynamic topology. The significant advancements in sensor, perception and onboard computation technologies has resulted in the rapid development of the vehicular network aiming at increasing human life quality by decreasing the number of traffic accidents and reducing the traffic jams. Over the past few years VANETs have demonstrated their significant potential in the development of Intelligent Transportation Systems (ITS) as well as traffic management systems and green transportation [24]. VANEts can provide applications across three categories, such as: Infotainment (including entertainment, navigation and telecom), traffic control with the main aim to reduce congestion and fuel consumption, and safety (including adaptive cruise control, forward collision warning, and speed regulation).

VANETs rely on Vehicle-to-Everything (V2X) communication systems that consist of Vehicle-to-Vehicle (V2V), Vehicle-to-Network (V2N), Vehicle-to-Infrastructure (V2I), and Vehicle-to-Pedestrian (V2P). Apart from the communication links, the VANET architecture consists of the On-Board Unit (OBU) which can have multiple different network interfaces and the Road Side Unit (RSU) that is the AP for vehicular communication. The two key enabling RATs that can enable V2X communications are the Dedicated Short-Range Communications (DSRC) and Cellular-V2X (C-V2X).

6.3.1 Wireless-Based Vehicular Communications

DSRC relies on several standards: the IEEE 802.11p amendment for wireless access in vehicular environments (WAVE), the IEEE 1609.2, 1609.3, and 1609.4 standards for Security, Network Services and Multi-Channel Operation, respectively. The **IEEE 802.11p** standard is used for the physical layer (PHY) and the medium access control (MAC) layer and it operates in the 5.9 GHz band. The standard makes use of seven 10 MHz channels in the 5.9 GHz band out of which four channels are used as service channels (SCH), one channel is reserved for future high availability low latency (HALL), one channel is used as control channel (CCH) and one channel is left unused. The CCH is used for safety messages and network control while the SCH is usef for all the other messages. All communication vehicles use CCH and one or ore SCH. The PHY is based on a variation of the IEEE 802.11a standard and makes use of OFDM with 64 sub-carriers used in 10 MHz and a data rate of up to 27 Mbps. The use of dedicated frequency band allows for less co-channel interference while the re-order of sub-carriers as compared to 802.11a enables a better multi-path mitigation. Additionally, 802.11p uses a longer guard period that can achieve a less inter-symbol interference and a better resistance against multi-path error. DSRC uses CSMA as the MAC protocol without the exponential back-off in order to reduce the latency that could be caused by a large contention window.

A set of stations within the IEEE 802.11p network form the WAVE Basic Service Set (WBSS) and the neighbouring WBSS use different channels. The WBSS can be formed by any WAVE device that wants to request an application. In general the devices that start the WBSS (OBU or RSU) are called the *provider* and they generate announcements. The *users* are the WAVE devices that join the WBSS. Two types of WBSS exists, such as: persistent WBSS - that are periodically announced with every synchronization interval, and non-persistent WBSS – that are announced on formation only and are short lived. When there are no active applications the WBSS is shutdown.

The WAVE devices can also communicate in the Outside the Context of a BSS (OCB) mode as they do not have to be a member of a WBSS to transmit. Consequently, a WAVE device can send a WAVE Short Message Protocol (WSMP) message to a broadcast address on CCH and other devices can respond to this message on the CCH. In this case there is no WBSS advertisement or synchronization required. However, in the OCB mode, the stations use a slightly higher Arbitration Inter-Frame Spacing (AIFS) than WBSS members as they have a lower priority.

The recent **IEEE 802.11bd** represents an evolution from IEEE 802.11p standard and integrates all the improvements brought from IEEE 802.11ax to enable the next generation of V2X communications. Some of the main goals of IEEE 802.11bd are as follows [25]: (1) enable backward compatibility and interoperability with IEEE 802.11p; (2) enable two times the throughput at MAC layer with a relative velocity of 500 kmph as compared to 802.11p; (3)

reduce packet collision and improve QoS under high Dopples shifts for improved reliability; (4) enable double the communication range as compared to IEEE 802.11p. All these goals can be achieved by the IEEE 802.11bd standard by integrating the following technical advancements [25]: 20 MHz channel bandwidth, support of higher rate MCS, up to 256QAM, 52 data sub-carriers, dual carrier modulation which enables the transmission of the same symbol twice over far-apart sub-carriers, adaptive re-transmissions where the re-transmission is decided based on the current congestion level, enable MIMO and 60 GHz mmWAVE support as well as localization improvements.

6.3.2   Cellular-Based Vehicular Communication

On the cellular networks side, the 3 GPP has developed C-V2X which is a LTE-A based RAT that leverages from the existing widespread cellular infrastructure to enable V2X communication. Consequently, in LTE-A the C-V2X communication can be enabled either over the traditional LTE air interface Uu (the interface between the eNodeB and the UE) operating in the mobile operator's licensed spectrum, as well as by using the sidelink air interface (PC5) that enables the direct communication between UEs without passing the data transmission through the eNodeB, and operating in the 5.9 GHz band.

Similarly to LTE, C-V2X is using the SC-FDMA and supports 10 and 20 MHz channels at 5.9 GHz and makes use of 180 kHz resource blocks (RBs) consisting of 12 sub-carriers of 15kHz each and occupying 0.5ms in time [26]. Transport Blocks (TBs) are used to transmit data over the Physical Sidelink Shared Channels (PSSCH) while the control information is packed into Sidelink Control Information (SCI) messages and sent over the Physical Sidelink Control Channels (PSCCH). PSSCH can be transmitted with QPSK or 16QAM according to the specified Modulation and Coding Scheme (MCS) while the control information over PSCCH is QPSK modulated.

C-V2X is designed for both in-coverage and out-of-coverage scenarios through two transmission modes, transmission mode 3 and transmission mode 4, respectively. Basically, the C-V2X sidelink mode 3 is defined for in-coverage scenarios and eNodeB is responsible of the scheduling of resources for vehicle communication. On the other side, the C-V2X sidelink mode 4 is defined for out-of-coverage scenarios where the vehicles UE autonomously perform resource selection without the involvement of the network. This is done based on the sensing of the environment. Transmission mode 4 is the one preferred by industry.

C-V2X makes use of semi-persistent scheduling of resources where the resources are assigned not only for the current transmission but for a number of subsequent transmissions [27]. In this way the scheduling overhead is reduced. Frequency reuse over a given geographical area is used in C-V2X mode 4. However, the increase in traffic density will reduce the frequency re-use distance which can result in increased interference level among the C-V2X users.

Currently, the 3GPP is working towards the development of New Radio (NR) V2X building atop of 5G NR that aims to work alongside C-V2X to enable the support of the use cases that cannot be supported by C-V2X. NR V2X is expected to have an enhanced sidelink design to support advanced V2X applications that require much more stringent QoS guarantees, enable coexistence of C-V2X and NR V2X within a single device, enable QoS management to meet the QoS requirements of different radio interfaces, Uu interface enhancements [27]. Apart from the C-V2X basic safety applications that NR V2X needs to support, it also needs to support some of the advanced V2X applications which require the end-to-end latency to be as low as 3 msec with reliability of 99:999% . NR V2X also defines sidelink mode 1 to enable direct vehicular communications within gNB coverage who also does the resource allocation for the UEs and NR V2X sidelink mode 2 that supports direct vehicular communications in the out-of-coverage scenario.

6.4 Millimeter Wave Multi-Gigabit Wireless Networks

One of the most promising frequency bands for the next generation multi-gigabit wireless networks is the millimeter Wave (mmWave) that makes use of a large unlicensed spectrum within the 30 GHz to 300 GHz bands. However, every country has its own frequency allocation and in general, manufacturers need to check the frequency bands when producing their products to see what countries they could cover in a certain frequency.

One of the Gigabit Wi-Fi amendment is the **IEEE 802.11ad** which operates in the 60 GHz band also referred to as the mmWave. The advantage is that there are fewer devices operating in this band, thus less interference. However, there is higher free space loss due to higher frequency band with poor penetration of objects which limits the usefulness of the standard to operate within one room only. Which in turn makes it more secure. The IEEE 802.11ad makes use of adaptive beamforming and high gain directional antennas with the devices being able to adapt their beams to an alternate path or finding paths off reflections from walls or other objects. It can support data rates of up to 7 Gbps making it suitable as a replacement for wires for video to TVs and projectors, high speed file transfer, interactive gaming, etc. Even though the mmWave at the 60 GHz band can bring many advantages there are several disadvantages that need to be considered as well, such as: *large attenuation* – due to the strong absorption by Oxygen which leads to short coverage areas, to address this it would require the use of larger transmit power and higher antenna gain for directional antennas; *directional deafness* – the devices cannot communicate unless they are aligned, carrier sense is not possible, RTS/CTS does not work and multicast is difficult; *easily blocked* –

the signal can be blocked even by a dog or a human which means that it requires relays.

The group of IEEE 802.11ad stations that communicate form a *Personal Basic Service Set* within which a PBSS Central Point (PCP) provides the scheduling and timing using beacons and performs antenna training with its members. The aim of the antenna training is for each station to find the optimal antenna configuration with its recipient using a two stage search, as follows: first there is a *Sector Level Sweep (SLS)* which aims to find the optimal sector for communication, and second the *Beam Refinement Procedure (BRP)* takes place which aims to search though the optimal sector to find the optimal parameters in that sector. Thus, through the antenna training there is a coarse finding and fine tuning happening in the sense that first the sector is found after that the optimal parameters within that sector. This process makes use of beam search which involves a binary search through sectors using beam steering and beam tracking where bits are appended to each frame to ensure that the beams are still aligned.

The next evolution of the IEEE 802.11ad standard is the **IEEE 802.11ay** which aims to support a maximum throughput of at least 20 Gbps, while maintaining or improving the power efficiency per station. This is achieved through a series of improvements at the PHY and MAC as compared to the 802.11ad including the use of aggregated channels, improved channel access and allocation for multiple channels, more efficient beamform training and beam tracking, Single-User Multiple-Input-Multiple-Output (SU-MIMO) and Multi-User Multiple-Input-Multiple-Output (MU-MIMO) operations [28].

6.5 Use-Case Scenarios: Trends in Heterogeneous Environments Integration

We see that the future in advancements in technologies is driven by the integration and interoperability of various technologies that come together under a heterogeneous environment to enable the best Quality of Experience (QoE) for the mobile users under different industries. This section aims to present the vision of three frameworks that bring together various technologies in different areas, such as:

- **5GonWheels** – a unified framework for reliable communication over heterogeneous vehicular networks;

- **I-RED** – an intelligent resource distribution and communications optimization framework within complex and dynamic networks like 5G-V2X;

- **PLATON** – Platform for Collaborative Learning And Teaching to Improve the Quality of Education Through Disruptive Digital Technologies;

6.5.1 5GonWheels

The latest developments in both automotive and communications industries, especially related to the current 5G networks, Internet of Vehicles (IoV) and adoption of V2X connectivity, are fuelling significant transformations in terms of driving Quality of Experience (QoE). The advent of the 5G-V2X paradigm aims at enabling effective connected cars communication as well as fully automated driving that could increase road safety and improve traffic management. However, in order to enable support for these services as well as a new set of related applications (e.g., traffic prediction, intelligent navigation systems, cooperative collision avoidance systems, etc.) one of the key requirements is provisioning of Ultra-Reliable and Low Latency Communications (URLLC) which cannot be guaranteed by the current underlying networks alone.

The highly dynamic nature of the vehicular networks along with the heterogeneity of wireless infrastructures for connected cars (e.g., IEEE 802.11p, LTE-A, etc.) as well as the variety of vehicular applications (e.g., safety, traffic management, infotainment, etc.) makes the resource management and the low latency communication requirements a significant challenge that cannot be handled by traditional networking solutions. In order to address these challenges many different solutions have been proposed in the research literature.

For example, Software-Defined Networking (SDN) emerged as an appealing solution also for vehicular environments as it allows controlling the network infrastructure in a centralized and programmable manner. Thus, many researchers looked into the integration of SDN in the context of VANET [29–31]. Ku et al. [29] proposed an adapted version of SDN to VANET environments by integrating three SDN components, such as: SDN controller which is the logical central intelligence, SDN wireless nodes which are SDN-enabled vehicles and SDN RSUs which are the SDN-enabled RSUs deployed along the road segments. Simulation results demonstrate the feasibility of the proposed solution in terms of SDN routing, fallback mechanism and transmission power adjustment. A similar approach was adopted in [30] where the authors propose a vehicular RSU cloud architecture consisting of traditional RSUs as well as SDN-based RSUs. The authors design a RSU cloud resource manager aiming at minimizing the delay, the operational cost and network reconfiguration. However, Zheng et al. in [31] argue that the use of only one SDN controller is not suitable for VANET as the underlying network topology is unstable due to the vehicles' high mobility. Thus, the authors propose a hierarchical control layer consisting of primary controller which maintains the global view of the network while secondary controllers enable QoS provisioning to low latency safety applications. Another promising solution to further reduce the latency within the VANET environment is the integration of Multi-access Edge Computing (MEC) [32]. In this context, the computation tasks of the vehicular applications are offloaded to MEC servers. Zhang et al. [32] propose an effective predictive scheme where the computation tasks are adaptively off-loaded to the MEC servers through direct uploading or predictive relay transmissions.

Numerical results indicate that the proposed scheme reduces the cost of computation while improving the task transmission efficiency. SDN could actually complement the use of MEC within VANET by enabling an agile solution and orchestrating the MEC edges across heterogeneous radio access technologies [33]. Al-Badarneh et al. [34] proposed a VANET-based Software-Defined Edge Computing infrastructure to enable the delivery of competitive services with reduced-latency. This is achieved by integrating vehicle-level caching for V2V communications as well as incorporating MEC capabilities inside network base stations. Basic evaluation results were obtained through Mininet-Wi-Fi emulator showing the benefits of the proposed approach in terms of latency.

Another main technology that has the potential to fulfil the URLLC requirements is Network Slicing (NSL). NSL consists of splitting a shared physical network infrastructure into multiple logical networks, referred to as slices. These slices are controlled and managed independently by the slice owners (i.e., OTT service providers or Virtual Mobile Network Operators (VMNO)) and can be used by one or multiple tenants (i.e., users). Each slice is allocated a set of network functionalities that are used only by that slice (the isolation concept) and which are selected from the shared network infrastructure. For example, one slice will be dedicated to collecting real-time data traffic while the other will be devoted to infotainment applications (i.e., Internet access). However, given the dynamism and scalability that slicing brings, managing and orchestrating these slices is not straightforward. Various approaches have been proposed to cope with these challenges. Kuklinski et al. [35] proposed DASMO, a distributed autonomous slice management and orchestration framework that addresses the management scalability problem. It uses the ETSI MANO combined with in-slice management approach to efficiently manage and orchestrate systems with a number of slices. Oladejo et al. [36] proposed a mathematical model to efficiently allocate radio resources in a two-level hierarchical network considering transmit power and allocated bandwidth constraints. The model is based on prioritizing network slices to cater for different users in a multi-tenancy network. The priority of each slice is determined by VMNOs based on which resources are allocated to the different slice while guaranteeing the minimum requirements per slice to ensure user satisfaction. The use of NSL with SDN was investigated by Costanzo et al. [37] by proposing a network slicing solution that makes use of an efficient scheduling algorithm, based on the centralized SDN architecture, to ensure a real-time allocation of bandwidth to the different slices considering their requirements in cloud radio access networks. Campolo et al. [38] proposed a theoretical framework consisting of a set of network slices representing different V2X use case categories. The framework makes use of NFV and SDN to provide different functionalities while meeting the requirements of various applications. One of these slices is dedicated to autonomous driving and it relies on V2V communication mode as the radio access technology and uses the core network to provide functions such as authentication, authorization and subscription management.

Combining NSL and MEC solutions in the context of vehicular networks and to be able to accommodate various vehicular applications with different QoS requirements, the network slices need to combine the available pool of network resources as well as cloud resources which are represented by the MEC utilities. Additionally, to enable the service customization in network slicing, the integration of SDN is essential as it will allow a dynamic and flexible service control [33]. Despite the amount of research done in this area, the lack of practical implementations of SDN deployments into challenging environments like MEC and VANET remains an open challenge [39]. Thus, the vision of 5GonWheels represents a unified framework integrating SDN, MEC and NSL in order to enable reliable communication over Heterogeneous Vehicular Networks (HetVNets). 5GonWheels is an innovative response to one of the main challenges of 5G related to URLLC and is designed to support low latency access to a wide range of safety and non-safety applications.

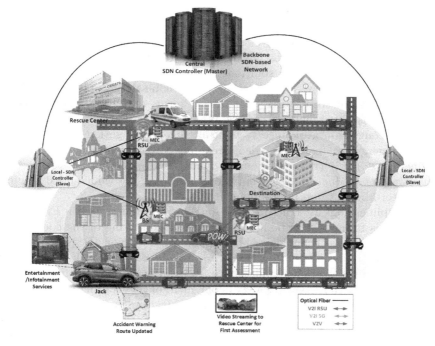

FIGURE 6.8
5GonWheels Vision

The main idea of 5GonWheels can be exposed using a use case scenario as illustrated in Figure 6.8. The scenario is inspired from the daily life of a commuter, named Jack, who every day before driving to his work place will drop his kids to school. Jack's car is a connected vehicle being connected to the 5GonWheels infrastructure. Initially, when starting from home Jack's car navigation system makes use of the 5GonWheels traffic management system to obtain the best route to destination while the Entertainment services will enable the kids to watch educational videos on their way to School. However, a traffic accident takes place on their path to destination which triggers the 5GonWheels infrastructure to define and enable a congestion area alert and disseminate it to the vehicles using V2V and V2I communication. Jack's vehicle will receive the accident warning message that will trigger the route update to avoid the congested area. Additionally, the 5GonWheels infrastructure will trigger the fast traffic accident rescue mission, which involves disseminating accident-related information from the involved vehicles sensors as well as real-time video streaming of the accident to the rescue center for a reliable initial assessment of the severity of the accident. This will enable the rescue center to prepare an efficient rescue by sending the appropriate equipment and expertize to the crash site and increasing the survivability chances of the injured people. Moreover, the 5GonWheels infrastructure will enable a fast rescue route for the emergency vehicle which involves prioritizing the emergency traffic by setting a dedicated shortest path as well as proactively informing the vehicles on that route to make room or clear the area for the rescue vehicle.

The above scenario is possible with the use of 5GonWheels framework, which represents a suite of solutions and integrated technologies enabling support for low latency access for safety and non-safety applications within HetVNets. As illustrated in Figure 6.8, in terms of vehicular applications, we can identify three main categories: (1) traffic safety applications (e.g., cooperative congestion avoidance system, intelligent navigation system in case of road emergency, real-time video streaming of emergency site for first assessment, etc.); (2) traffic efficiency applications (e.g., shortest route, time to destination, reduction of gas emissions, etc.) and (3) value-added services (e.g., entertainment and infotainment).

A simplified logical structure of 5GonWheels is illustrated in Figure 6.9. 5GonWheels framework is built over HetVNets consisting of the coexistence of different wireless technologies like cellular Base Station (BS)s (e.g., LTE) and RSUs and OBUs based on DSRC to provide V2I and V2V connectivity and overcome the sporadic connectivity issues of highly mobile and dynamic environments. Because of the widely deployment of the LTE network, it makes it a promising solution to enable V2I communication. Another option is the use of DSRC systems which are designed to provide robust, low-latency, and high-throughput services for V2I and V2V communications making them suitable for safety and non-safety applications. However, because of their sparse deployment and short range they provide intermittent connectivity. Thus, the coexistence of these two technologies creates a HetVNets environment able to provide a continuous vehicular connectivity and enables the support for Connected and Automated Vehicle (CAV)s. 5GonWheels framework is built on top of three main technologies: SDN, MEC and NSL. In order to address the problems of interoperability and interconnectivity of heterogeneous radio technologies, SDN is employed. SDN separates the control plane from the

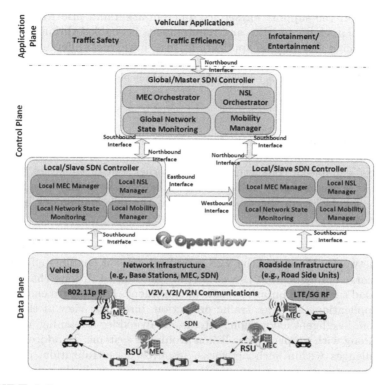

FIGURE 6.9
5GonWheels Framework

data plane and provides a centralized control of the network infrastructure. However, in the context of HetVNets, a fully centralized SDN solution relying on one unique controller only would not be suitable due to the large-scale and highly dynamic nature of the vehicular networks. This could lead to a significant increase in the control traffic which will cause increase in latency and network response time [31]. Thus, in order to fit the ultra-low latency requirements of the future vehicular 5G networks, 5GonWheels adopts a hierarchical organization of the controllers as illustrated in Figure 6.9, where a centralized master controller maintains the global view of the network, while local slave controllers are used to manage strict QoS-requirements applications. Additionally, to further reduce the latency and increase the network response time, a hybrid approach is considered where MEC is deployed at the BSs and RSUs. In this way, the local SDN controllers will transfer some computation tasks and services to the nearby MEC servers based on their local knowledge. The integration of MEC will enable 5GonWheels to achieve low latency access to safety and non-safety services.

Moreover, the support for service differentiation and prioritization based on application characteristics and requirements is enabled through the use of network slicing. As various applications may have very different and highly dynamic requirements in terms of latency, reliability, etc. an optimal dynamic network slicing solution is employed in 5GonWheels.

6.5.2   I-RED

The growing ubiquity of mobile devices and services' driven applications has transformed the communications networks and has led to a significant number of 'things' capable of connecting to the Internet, out of which an important portion is represented by vehicles. This, together with the ever-increasing number of vehicles on the roads has led to the development of Internet of Vehicles (IoV).

Interconnecting billions of IP enabled devices as well as the adoption of Vehicle-to-Everything (V2X) connectivity give rise to truly complex and dynamic networks. Moreover, the V2X paradigm aims at enabling effective connected cars communication as well as fully automated driving that could increase road safety and improve traffic management. This in turn enables a new set of applications, such as: traffic prediction, bird's eye view at the road intersection, intelligent navigation systems, 4K live video streaming, augmented reality along with cooperative collision avoidance systems. To address the critical challenges within such networks, resource decentralization, distributed computing, storage and communications can play a vital role.

In this context, I-RED is an intelligent resource distribution and communications optimization framework within complex and dynamic networks like 5G-V2X. I-RED makes use of smart distributed computing, communications and resource optimization to improve users' QoE for value-added services (e.g. infortainment, business), improved driving experience for transportation efficiency (e.g., reduction of gas emissions, time to destination, congestion control, smart navigation), as well as the users' safety for traffic safety applications (e.g., cooperative collision avoidance system, intelligent navigation system in road emergency and developing hazards, etc.). I-RED targets efficient resource distribution, multi-threaded analysis and service migration to mitigate the limitations of existing cloud services, vehicular networks, cellular systems, and emergency communications.

One of the key requirements for the emerging 5G-V2X is the provisioning of ultra reliable low latency communications which cannot be guaranteed by the current underlying networks. Thus, I-RED aims to address the challenges arising in highly dynamic networks in the context of 5G-V2X in three key areas: (1) Edge and Fog Computing - I-RED introduces appropriate mechanism to distribute data and tasks among the fog and edge devices to establish efficient work flow and improved application performance. (2) Resource Optimization – I-RED targets intelligent resource sharing to improve user experience and service credibility and investigates the suitable programming models to implement context aware functionality and operation transparency. In addition, contextual data flow, information sharing and storage will be improved to minimize channel congestion, network load, resource usage and communications optimization in 5G-V2X. (3) Traffic Classes and Applications Prioritization – the development of effective distribution systems, resource handling, computation and communications will result in appropriate identification of application classes and realization of platforms along with performance bencharking in ICT and its extended applications.

In recent years, the number of connected devices has increased significantly[40]. Additionally, the bandwidth and resource requirements for a single device have seen a notable rise due to the incorporation of smart services (infotainment, educational, marketing, business, safety, transportation). Interconnecting diverse applications (e.g. autonomous transportation, safety, emergency, infotainment etc.) and technologies (e.g. LTE, 5G, V2X, fog/edge computing) gives rise to truly complex and dynamic networks. The incorporation of high mobility users with strict time and reliability requirements in ever changing network dynamics (with real time connection formation/deformation) also introduce unique challenges. The evolved capabilities of 5G, V2X and distributed computing must support the future needs of ultra-reliable and low latency communications, increased network bandwidth, massive number of devices, smart distributed processing and timely executions and decisions [41].

Several development plans targeting 5G systems, IoT, Connected Autonomous Vehicles (CAV), Massive Machine Communications (MMC), smart sensors, machine intelligence and cloud services are underway to propose suitable frameworks to address the challenges in highly diverse and dynamic networks [41–44].

Unfortunately, the developments in distributed computing frameworks is not well investigated in continuously evolving and ever changing dynamic networks. The existing ICT forms a complex, non-flexible and disjoint solution targeting limited areas of applications [42]. Most of the research focuses on very limited application aspects and lack practical implementation. In addition, the intelligent resource distribution and distributed computing is not addressed appropriately in existing studies [42; 43].

Real-time intelligent decisions, network optimization and support for transportation, safety and emergency applications, require high computing and communications resources, effectively managed through edge devices. The developments in efficient resource management, parallel programming, congestion control, communications optimization, service migration and prioritized access can address several prevailing challenges in distributed computing and communication networks. In addition, these developments can support reliable decisions in critical applications (emissions control, traffic congestion control, smart navigation, cooperative collision avoidance and road emergency responsiveness, etc.). Therefore, development of suitable solutions to integrate diverse computing and communications resources is necessary to cope with ever rising requirements of immensely expanding networks.

Fog computing serves as an intermediate transitional layer between the cloud and end-user devices where fog offers reduced access delay and network congestion by localizing the desired content in the neighbourhood of the end user devices. The cloud computing serves as a central pool of resources which offers rich (on-demand) services such as computing and storage. Although, cloud computing offers several benefits yet, is located far away from the end users which results in unwanted delay. The Fog computing is helpful in minimizing cloud limitations. It keeps localized information and services to provide easy access for end users. Fog serves as a local cloud to meet the real-time requirements of end-users. There may be many Fog devices including, servers and interconnected nodes to provide low latency services. The Fog servers collect localized information from the cloud to optimize information flow and timely data provision for end users whenever required. When any end user in the vicinity of that Fog server requests the information; it is provided without the added delay caused in fetching data from the cloud. Fog servers and nodes may communicate with each other and other mobile devices at user's end. The end user layer contains a variety of mobile devices looking for different services from the Fog. This layer is also called "edge" where end users having diverse interests are supposed to access the services provided by the Fog. These mobile devices can connect to the nearest Fog server for the localized information. Mobile devices at the edge may spread over multiple locations and be connected with the respective Fog server which is ultimately connected with the Cloud.

To facilitate effective communications, the fog devices must proactively devise strategies to share computational as well as communications resources. In addition, the diverse user services/applications must also be prioritized for improved user experience and quality of life. I-RED focuses on application prioritization based on three main types of vehicular traffic classes identified, such as: (1) value-added services (e.g., infotainment, business); (2) traffic efficiency applications (e.g., reduction of gas emissions, time to destination, etc.); (3) traffic safety applications (e.g., cooperative collision avoidance system, intelligent navigation system in case of road emergency, etc.). The I-RED framework is primarily integrating diverse services including 5G, LTE, CAV, IoT, Fog and cloud. Therefore, interlinking different services and technologies is of utmost importance. The transition between these resources and optimization not only requires high computational capabilities but also need parallel processing for timely decision. Therefore, I-RED will proactively distribute the computational tasks among the variety of edge and fog devices in order to achieve ultra reliable low latency communication, which is one of the key requirements for the future 5G-V2X. Distributive multi-threaded algorithms will be developed to receive timely evaluation of optimal resource allocation and priority based services for individual users. In addition, the overall network congestion will also be minimized in state of the art ICT with localized processing and control. A graphical representation of such system handling diverse services is presented in Figure 6.10.

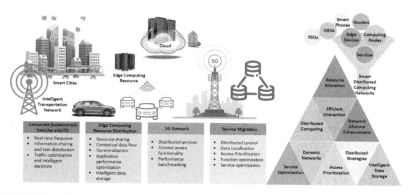

FIGURE 6.10
Smart Distribution of Resources in Dynamic Networks

I-RED proposes application prioritization techniques that target the following challenges: (1) improving Quality of Experience for value-added services (2) improving Quality of Driving Experience (3) improving Quality of Safety Living through road. I-RED makes use of the integration of edge and fog computing to propose new service migration techniques for intelligent data storage, processing and movement.

The effectiveness of CAV, integrated services and prioritized access for emergency situations can significantly optimize the emergency response services. The critical nature of different emergency calls whether it is due to a road accident, severe deterioration in health signs of a regular citizen or a natural disaster, in all such cases the first few minutes are termed very important and can govern the decision of someone's life and death. For these purposes the emergency response staff and ambulances are equipped with state of the art medical equipment to provide standard first aid response to the critical patients during transportation. However, the skill level of the emergency response team is relatively limited compared to the hospital staff. In such cases, technology assisted supervision of highly qualified surgeons, remotely connected to the ambulance feed, can improve the overall survival chances of the critical patients. Furthermore, the present-day technology is also capable of forming ad-hoc ambulance vehicles network, which, based on the location of different vehicles in the network, can suggest most suitable and accessible vehicle to the emergency call location. The emergency response vehicle selection algorithm can also consider the real-time traffic parameters for suitable route selection. Furthermore, the real-time feed from the ambulance can give a better chance to the medical staff at the hospital to prepare beforehand for the arriving patient and treat him/her more effectively on arrival.

Since the basic requirements for the proposed systems need a roadside infrastructure which offer effective road communication and in-vehicle intranet of Internet Enabled Devices (IEDs), capable of offering video feed and information of vital signs including blood pressure, heart rate, respiratory rate, ECG trace etc. However, to analyse the real-time traffic details and road blocks need to be handled on larger scale which do bring the Fog devices in play. Fog, with the help of edge devices, can devise suitable route plan using collected information from the area of interest. The critical information communications and localized distributed computation for optimal route selection can play a very important role. Furthermore, service migration along the path for continuous video feed, must enable seamless transition between different technologies and BSs.

6.5.3 PLATON

The COVID-19 global pandemic crisis has significantly disrupted the higher education sector with temporary closure that forced the educational institutions to rapidly embrace technology-enhanced learning. Consequently, the COVID-19 containment measures that forced people to work or stay at home led to the rapid digitalization of the education that moved to the online world, with the educational content being delivered remotely using technology. However, in this context we can identify several factors that could have an impact on the success of moving towards remote learning, such as the underlying digital maturity of the pre-existing Information and Communications Technology (ICT) infrastructure and systems; the requisite supporting skills and digital literacy among the students and instructors; unequal wealth and unequal access to technology among students; the nature of different subjects where practical skills are essential, etc. Until now, digitalization was seen as an opportunity to eradicate inequity in education. However, the above mentioned challenges reflect that when it comes to implementing technology-enhanced learning approaches the one-size-fits-all approach must be avoided and the adopted strategies have to suit the needs of all, to avoid exacerbating the pre-existing inequalities even further. During the 2021 spring annual meeting of the United Nations Broadband Commission for Sustainable Development[45], it was highlighted the importance of global broadband connectivity for everyone everywhere in order to tackle the educational inequalities and build an inclusive post-COVID digital future for all. To achieve this, there is a need for personalized digital solutions and services that could take into consideration the needs of the users and their digital disparity and ensure digital equity.

Apart from the societal and cultural changes the world is facing due to COVID-19 pandemic, the changes in customer behaviour is also pushing the major industries to reconfigure their processes and business. Consequently, we are dealing with a new workplace world with digital-first, contactless operations, remote access, physical distancing, etc. that along with automation and digitisation have a significant impact on the current workforce skills and capabilities, leading to significant skill gaps [46]. Moreover, the European Union highlights the necessity of having educational and training systems that are fit for the current digital age and has issued the Digital Education Action

Plan (2021-2027) [47] that sets out two priority areas: (1) fostering the development of high-performing digital education ecosystem and (2) enhancing digital skills and competences for the digital transformation.

In this context, key enabling technologies, such as Multi-access Edge Computing (MEC), Artificial Intelligence (AI) and Machine Learning (ML), rapid deployment of next generation networks (5G) have accelerated the adoption of on-demand remote learning across diverse industries. This in turn has led to the e-learning market being projected to exceed USD 375 Billion by 2026 [48]. Consequently, there are many efforts looking at leveraging technology in education for digital transformation and drive the industry demand for e-learning and remote/distance education. In this regards, Augmented reality (AR)/virtual reality (VR)/ mixed reality (MR) are key enablers for efficient e-learning solutions [49].

Different solutions have been proposed in the literature, that are trying to improve the digital experience in education through advanced multimedia applications delivered over current and next generation mobile networks [50]. It is well known that some of the most promising approaches that could cope with the dynamics of the underlying wireless networks are the adaptive multimedia solutions [51]. Various industries already adopted commercial adaptive streaming solutions like Apple's HTTP Live Streaming (HLS), Microsoft's Silverlight or even the more recent security-enhanced adaptive solution proposed by Akamai [52]. These solutions help increase the users' Quality of Experience (QoE) in general. However, research studies have shown that within an educational context, the adaptive multimedia solutions have the potential to increase learners QoE and improve academic performance [53]. The diversity of adaptive multimedia delivery solutions proposed in the literature range from server-located adaptive decision-making solutions [54], client-located DASH-based approaches [55], schemes that go beyond conventional multimedia adaptation by targeting multiple sensorial content delivery [56] or omnidirectional video [57] and finally generic adaptive solutions [58], extended adaptive mechanisms based on saliency information gathered using eye tracking [59], or network characteristics-aware resource scheduling solutions to enable remote education [60]. Moreover, EU funded a number of research projects (e.g., HandiCAMs, QoSTREAM, CONCERTO, etc.) focused on multimedia adaptation techniques to increase the end users' QoE.

In order to achieve an improved QoE level and increase in the academic performance for the learners, the proposed solutions are becoming increasingly complex especially when applied within a heterogeneous environment. Consequently, there is a need for innovative approaches such as the use of AI/ML in education. Due to the advances in technology there is a shift towards promoting technological innovation in education such as the integration of AI to enable support for informed, data-driven decision making [61].

Recently, there has been a significant increase in the usage of AI-based online learning applications to support remote learning [62]. This could represent the start of Education 5.0 revolution, where AI-enabled digital learning platforms could guide students on how they learn and how to acquire the required skills regardless of their socio-economic background, geographical location or gender [63]. However, the adoption of AI brings some new challenges along, such as: the fairness of the AI algorithms, the proneness to bias, its interpretability, and its appropriate and ethical use [59]. Consequently, previous studies have shown that AI in education needs interpretable and Explainable Machine Learning (XML) [64]. The ability to understand what is happening within a ML model and know exactly how it works to be able to describe it in human terms is referred to as XML [65]. In general, the ML algorithms operate as a black box, without access to the basis of the decisions outcome. Consequently, the traceability of the decision process is inhibited due to the high complexity of the ML algorithms limiting their explainability. Further research is required to understand how to integrate XML-based solutions within the educational context and make them viable in supporting the learners and instructors at scale.

Other disruptive technologies that gained recent popularity in education include collaborative VR environments [66]. These involve creation of a virtual learning environment that could prompt or guide efficient learning [67] and could actually be used as a tool to compensate for the lack of face-to-face interaction between teachers and learners while enabling the learners to further develop their practical skills [68]. However, these multimedia-rich intelligent educational environments are known to have strict Quality of Service (QoS) requirements. These requirements include ultra-high bandwidth, ultra-large storage and ultra-low latency that could pose significant challenges and could put tremendous pressure on the underlying networks. MEC is seen as one possible solution to overcome these challenges and drive the learners' QoE requirement for multimedia-rich applications within smart educational environments [69]. The introduction of MEC could enhance the QoE by bringing storing and computing services at the edge of the network, closer to the end user. MEC could be used in conjunction with adaptive network delivery of VR content to improve the service quality [70].

Despite the amount of research done in this area, there is a need to go beyond the basic requirements of access to a digital device and stable internet connectivity. There is a need to address the lack of a practical intelligent technology-based educational environment, which bridges the gaps between the avenues discussed. In this context, PLATON represents a unified framework integrating XML for personalized adaptive technology-enhanced learning, ML-based VR environment to enable practical skills and MEC to meet the stringent network requirements of such an technology-rich multimedia environment. PLATON is an innovative response to one of the main challenges of technology-enhanced learning related to the quality of education. PLATON is designed to avoid the one-size-fits-all approach and addresses learners' needs, regardless of their socio-economic background, geographical location or gender.

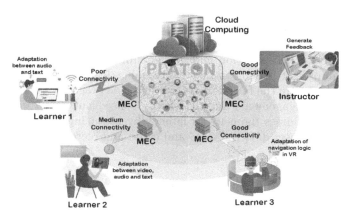

FIGURE 6.11
PLATON Vision

The main idea of PLATON can be presented using a use case scenario, illustrated in Figure 6.11. The scenario is inspired from the educational landscape during the global COVID-19 pandemic that forced the higher education institutions to move to online delivery. Figure 6.11 illustrates a use case scenario involving three learners and one instructor connected to the PLATON platform. The learners could be located at different geographical locations presumably all over the world, with access to different device types (smartphones, laptops, tables, VR gear, etc.) and different network connectivity characteristics (poor, medium, good connectivity, etc.). The learners are also diverse, they have unique cognitive characteristics and abilities and different learning styles. Taking into account all these challenges, PLATON aims to deliver personalized, adapted educational material (e.g., text, audio, video, VR, etc.) that is tailored to each learners' needs within a collaborative environment. For example, as illustrated in Figure 6.11, Learner 1 has a poor network connectivity which means that trying to access the video streaming would have a negative impact on the overall learner's experience. Consequently, based on the network connectivity and learner's profile PLATON will adapt the educational content (e.g., audio and text) delivered to the learner. Similarly, Learner 2 with a medium network connectivity would be able to receive personalized adaptive educational content (e.g., a mix of video, audio and text), while Learner 3 with good network connectivity could engage in immersive learning scenarios using the VR gear. The adaptation in this case is done at the level of navigation logic within the VR environment. XML is employed for personalization and adaptation of the educational content to enable the interaction of Learners having different learners' profiles, backgrounds, and requirements and the PLATON platform. PLATON will be able to generate multimodal explanations and quantify the degree of interpretability by revealing the "Why?", "Why not?", "What if?" within the decision process to the learner. Moreover, it is well known that video streaming is already pushing the limits of the underlying network infrastructure while the increasing popularity and adoption of VR applications will account for additional strains on the networks. VR adoption is of paramount importance for the future of education in order to enable immersive teaching/learning experiences for the learners and prepare the future workforce with the practical skills required for the real-world engagements and activities. Consequently, the integration of MEC will help address the network latency by balancing the data processing between the end user devices, MEC and cloud and increase the overall users' QoE.

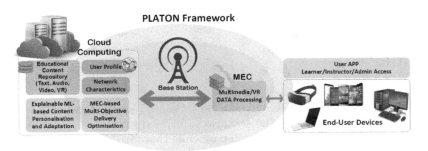

FIGURE 6.12
PLATON Framework

The above scenario is possible with the use of PLATON framework, which represents a suite of solutions and integrated key technologies enabling a collaborative learning and teaching environment to improve the learners' QoE through low latency access for personalized and adaptive educational content. As illustrated in Figure 6.11, PLATON contributes to advances of three main directions: (1) XML for personalized adaptive technology-enhanced learning; (2) enhancing practical skills through technology-rich multimedia environments (i.e., VR and ML); (3) MEC integration to reduce latency and improve learners' QoE.

A simplified logical structure of PLATON with the main functional block is illustrated in Figure 6.12. PLATON framework is built as a distributed solution, consisting of the coexistence of different key disruptive technologies. At the cloud side the following blocks are located: (1) educational content repository which stores the educational content in different formats, such as text, audio, video, VR; (2) user profile – stores information about the learners and their different ways of learning to model the learner profiles; (3) network characteristics – stores information about the current network conditions at the learner location; (4) XML-based content personalisation and adaptation – makes use of XML to personalize and adapt the educational content based on the learners' profiles and network conditions; (5) MEC-based multi-objective delivery optimisation – finds the best trade-off between cloud vs. MEC vs.

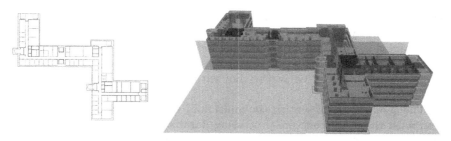

FIGURE 6.13
2D and 3D Geometry of a Multi-Floor Building

end user device computational workload in order to reduce the overall network latency and improve QoE. A multimedia/VR Data processing block is located at the MEC side while at the end-user side we have various types of devices and the user app with different modes of access, such as learner, instructor or admin. The user app will follow the digital accessibility requirements by making the PLATON platform perceivable, operable, understandable and robust to allow everyone, including people with disabilities to interact with the educational content.

6.6 Practical Use-Case Scenario: Wireless Indoor Communication Using Altair WinProp

The aim of this practical use-case scenario is to use WinProp in order to investigate the network planning and coverage area offered by a wireless local area network (e.g., IEEE 802.11g) in an indoor scenario of a multi floor building [71]. To be able to achieve this, we are going to use a detailed indoor geometry description of a multi floor building illustrated in Figure 6.13, with exterior walls, windows, interior walls made of different materials, doors, and other relevant items.

In this indoor environment, we are going to analyze the impact of three radio propagation models, namely multi-wall model COST 231, Intelligent Ray Tracing (IRT) and the Dominant Path Model (DPM). Consequently, we need to pre-process the *.idb* file provided by Altair [71] before using it in ProMan. For this a new project is created in WallMan (File > New Project) that will use the *.idb* file. We then use the Preprocessing > Edit Preprocessing Parameters option to select the mode for the pre-processing run for each propagation model. The output would be three different database files for each mode. Thus, the *.idc* database file represents the indoor COST 231 Binary database, the *.idi* database file represents the IRT Binary database and the *.idp database file represents the DPM Binary database. These files are generated by selecting *Preprocessing > ComputeCurrentProject* in the menu and they contain indoor building data as well as the visibility information along with the considered parameters for this pre-processing. These files are now ready to be used in ProMan with their corresponding propagation model.

The geometry representing the multi floor building was generated in Wall-Man with three different pre-processing database files, *.idc, *.idi and *.idp each corresponding to a propagation mode such as COST 231, IRT and the DPM, respectively. We will now use all these database files in the propagation simulation tool ProMan to produce coverage plots and study the difference in coverage area in terms of received power and maximum achievable throughput when using the three different radio propagation models within the same indoor environment and the same access points antennas characteristics deployed at the same locations.

The overall workflow in ProMan is as follow:

- load an Indoor Scenario

- load the IEEE 802.11g air interface specified by the *.wst file

- load the preprocessed vector database that was prepared in WallMan.

- define source locations.

- specify antenna characteristics.

- set up simulation, including selection of a radio propagation model.

- run simulation.

- inspect results.

The IEEE 802.11g air interface makes use of OFDM/SOFDMA for multiple access and the TDD mode for switching between uplink and downlink. A total number of eight sites with omni-directional antennas operating at 2.4 GHz and transmit power of 16 dBm were deployed throughout the multi floor building across four levels (e.g., ground floor to thrid floor) with two sites per floor. The predictions are computed at the following heights above ground level corresponding to the four floors, such as: 1.5 m, 5.2 m, 8.9 m and 12.6 m. In order to minimize the interference between the antennas, four carrier frequencies in the 2.4 GHz band were used among the antennas. The simulations were configured for homogeneous traffic per cell.

Figures 6.14, 6.15 and 6.16 represent the predicted received power per each floor of the multi floor building, for the three propagation models, COST 231, IRT and the DPM, respectively. The advantages of using the DPM model is that it focuses on the most relevant paths only, speeding up the computation process. The COST 231 model does not take into account the impact

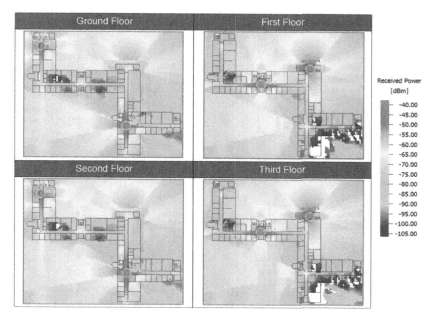

FIGURE 6.14
Predicted Received Power per Floor Using IRT

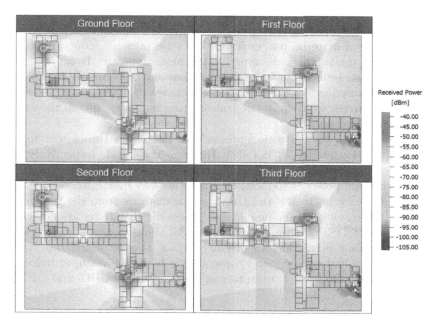

FIGURE 6.15
Predicted Received Power per Floor Using COST 231

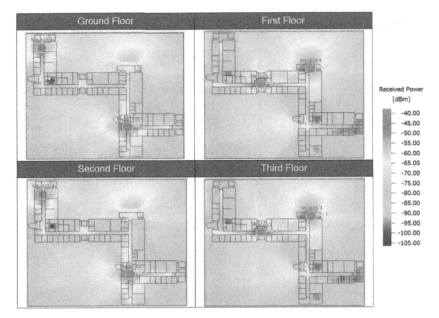

FIGURE 6.16
Predicted Received Power per Floor Using DPM

of diffraction and multipath propagation. However, the model considers the different types of material properties for the internal walls, doors, floors and ceilings which makes the indoor prediction more accurate. It can be noted that the internal walls around the transmitting antennas are obstructing the direct rays. Consequently, the IRT model becomes too pessimistic, especially at distances further away from the transmitting antennas, like in the case of the first and third floors.

Figures 6.17, 6.18 and 6.19 represent the maximum throughout for the three propagation models COST 231, IRT and the DPM, respectively. The white areas indicate that the communication is not possible due to the fact that the received power is too low. It can be noticed that in the case of first and third floor for both IRT and DPM, as the distance from the transmitter increases, the results become pessimistic. These disadvantages are not visible in the case of COST 231 model as this model shows comprehensive results across the indoor prediction area. Thus, it can also be noted that COST 231 can achieve higher power and consequently higher throughput indoor, at further distances from the transmitter as compared to bot IRT and DPM.

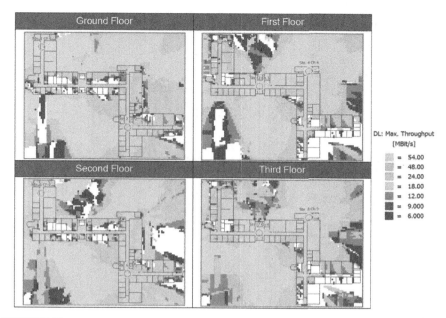

FIGURE 6.17
Maximum Throughput per Floor Using IRT

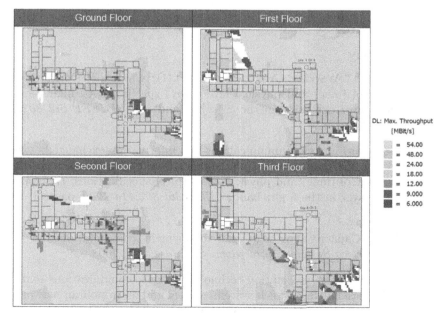

FIGURE 6.18
Maximum Throughput per Floor Using COST 231

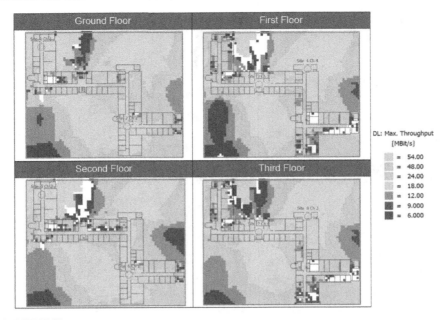

FIGURE 6.19
Maximum Throughput per Floor Using DPM

Bibliography

[1] IEEE 802.11 Working Group et al. Part 11: wireless lan medium access control (mac) and physical layer (phy) specifications: higher-speed physical layer extension in the 2.4 ghz band. *ANSI/IEEE Std 802.11*, 1999.

[2] Ieee standard for telecommunications and information exchange between systems – lan/man specific requirements – part 11: Wireless medium access control (mac) and physical layer (phy) specifications: High speed physical layer in the 5 ghz band. *IEEE Std 802.11a-1999*, pages 1–102, 1999.

[3] R Seide. Capacity, coverage, and deployment considerations for ieee 802.11 g. *cisco Systems white paper, San Jose, CA*, 2003.

[4] Ieee standard for information technology – telecommunications and information exchange between systems – local and metropolitan networks - specific requirements – part 11: Wireless lan medium access control (mac) and physical layer (phy) specifications: Higher speed physical layer (phy) extension in the 2.4 ghz band. *IEEE Std 802.11b-1999*, pages 1–96, 2000.

[5] Ieee standard for information technology – local and metropolitan area networks – specific requirements – part 11: Wireless lan medium access control (mac) and physical layer (phy) specifications: Further higher data rate extension in the 2.4 ghz band. *IEEE Std 802.11g-2003 (Amendment to IEEE Std 802.11, 1999 Edn. (Reaff 2003) as amended by IEEE Stds 802.11a-1999, 802.11b-1999, 802.11b-1999/Cor 1-2001, and 802.11d-2001)*, pages 1–104, 2003.

[6] Ieee standard for information technology – local and metropolitan area networks – specific requirements – part 11: Wireless lan medium access control (mac)and physical layer (phy) specifications amendment 5: Enhancements for higher throughput. *IEEE Std 802.11n-2009 (Amendment to IEEE Std 802.11-2007 as amended by IEEE Std 802.11k-2008, IEEE Std 802.11r-2008, IEEE Std 802.11y-2008, and IEEE Std 802.11w-2009)*, pages 1–565, 2009.

[7] Evgeny Khorov, Ilya Levitsky, and Ian F Akyildiz. Current status and directions of ieee 802.11 be, the future wi-fi 7. *IEEE access*, 8:88664–88688, 2020.

[8] Evgeny Khorov, Anton Kiryanov, Andrey Lyakhov, and Giuseppe Bianchi. A tutorial on ieee 802.11 ax high efficiency wlans. *IEEE Communications Surveys & Tutorials*, 21(1):197–216, 2018.

[9] Ieee standard for information technology – local and metropolitan area networks– specific requirements – part 11: Wireless lan medium access control (mac)and physical layer (phy) specifications amendment 1: Radio resource measurement of wireless lans. *IEEE Std 802.11k-2008 (Amendment to IEEE Std 802.11-2007)*, pages 1–244, 2008.

[10] Sven D Hermann, Marc Emmelmann, O Belaifa, and Adam Wolisz. Investigation of ieee 802.11 k-based access point coverage area and neighbor discovery. In *32nd IEEE Conference on Local Computer Networks (LCN 2007)*, pages 949–954. IEEE, 2007.

[11] Bernhard H Walke, Stefan Mangold, and Lars Berlemann. *IEEE 802 wireless systems: protocols, multi-hop mesh/relaying, performance and spectrum coexistence*. John Wiley & Sons, 2007.

[12] J Polk, John Schnizlein, and Marc Linsner. Dynamic host configuration protocol option for coordinate-based location configuration information. *Network Working Group, Request for Comments*, 3825, 2004.

[13] Ramona Trestian. *User-centric power-friendly quality-based network selection strategy for heterogeneous wireless environments*. PhD thesis, Dublin City University, 2012.

[14] William Stallings. *Wireless Communications & Networks: Pearson New International Edition PDF eBook*. Pearson Higher Ed, 2013.

[15] Behrouz A. Forouzan. *Data communications and networking*. McGraw-Hill, Inc., USA, 3 edition, 2003.

[16] Jochen H Schiller. *Mobile communications*. Pearson education, 2003.

[17] Alex Hinds, Michael Ngulube, Shaoying Zhu, and Hussain Al-Aqrabi. A review of routing protocols for mobile ad-hoc networks (manet). *International Journal of Information and Education Technology*, 3(1):1, 2013.

[18] Charles E Perkins and Pravin Bhagwat. Highly dynamic destination-sequenced distance-vector routing (dsdv) for mobile computers. *ACM SIGCOMM Computer Communication Review*, 24(4):234–244, 1994.

[19] Mehran Abolhasan, Tadeusz Wysocki, and Eryk Dutkiewicz. A review of routing protocols for mobile ad hoc networks. *Ad Hoc Networks*, 2(1):1–22, 2004.

[20] Shree Murthy and JJ Garcia-Luna-Aceves. A routing protocol for packet radio networks. In *Proceedings of the 1st annual international conference on Mobile computing and networking*, pages 86–95, 1995.

[21] David B Johnson and David A Maltz. Dynamic source routing in ad hoc wireless networks. In *Mobile computing*, pages 153–181. Springer, 1996.

[22] Charles Perkins, Elizabeth M Royer, and S Das. Ad-hoc on demand distance vector routing (aodv). Technical report, Internet-Draft, November 1997. draft-ietf-manet-aodv-00. txt, 2003.

[23] Nicklas Beijar. Zone routing protocol (zrp). *Networking Laboratory, Helsinki University of Technology, Finland*, 9(1):12, 2002.

[24] Irina Tal. *Improving User Experience and Energy Efficiency for Different Classes of Users in Vehicular Networks*. PhD thesis, Dublin City University, 2016.

[25] Xiaomin Ma and Kishor S Trivedi. Sinr-based analysis of ieee 802.11 p/bd broadcast vanets for safety services. *IEEE Transactions on Network and Service Management*, 18(3):2672–2686, 2021.

[26] Valerian Mannoni, Vincent Berg, Stefania Sesia, and Eric Perraud. A comparison of the v2x communication systems: Its-g5 and c-v2x. In *2019 IEEE 89th Vehicular Technology Conference (VTC2019-Spring)*, pages 1–5. IEEE, 2019.

[27] Gaurang Naik, Biplav Choudhury, and Jung-Min Park. Ieee 802.11 bd & 5g nr v2x: Evolution of radio access technologies for v2x communications. *IEEE access*, 7:70169–70184, 2019.

[28] Pei Zhou, Kaijun Cheng, Xiao Han, Xuming Fang, Yuguang Fang, Rong He, Yan Long, and Yanping Liu. Ieee 802.11 ay-based mmwave wlans: Design challenges and solutions. *IEEE Communications Surveys & Tutorials*, 20(3):1654–1681, 2018.

[29] Ian Ku, You Lu, Mario Gerla, Rafael L Gomes, Francesco Ongaro, and Eduardo Cerqueira. Towards software-defined vanet: Architecture and services. In *2014 13th Annual Mediterranean Ad Hoc Networking Workshop (MED-HOC-NET)*, pages 103–110. IEEE, 2014.

[30] Mohammad Ali Salahuddin, Ala Al-Fuqaha, and Mohsen Guizani. Software-defined networking for rsu clouds in support of the internet of vehicles. *IEEE Internet of Things Journal*, 2(2):133–144, 2014.

[31] Kan Zheng, Lu Hou, Hanlin Meng, Qiang Zheng, Ning Lu, and Lei Lei. Soft-defined heterogeneous vehicular network: Architecture and challenges. *IEEE Network*, 30(4):72–80, 2016.

[32] Ke Zhang, Yuming Mao, Supeng Leng, Yejun He, and Yan Zhang. Mobile-edge computing for vehicular networks: A promising network paradigm with predictive off-loading. *IEEE Vehicular Technology Magazine*, 12(2):36–44, 2017.

[33] Tarik Taleb, Konstantinos Samdanis, Badr Mada, Hannu Flinck, Sunny Dutta, and Dario Sabella. On multi-access edge computing: A survey of the emerging 5g network edge cloud architecture and orchestration. *IEEE Communications Surveys & Tutorials*, 19(3):1657–1681, 2017.

[34] Jafar Al-Badarneh, Yaser Jararweh, Mahmoud Al-Ayyoub, Ramon Fontes, Mohammad Al-Smadi, and Christian Rothenberg. Cooperative mobile edge computing system for vanet-based software-defined content delivery. *Computers & Electrical Engineering*, 71:388–397, 2018.

[35] Slawomir Kukliński and Lechosław Tomaszewski. Dasmo: A scalable approach to network slices management and orchestration. In *NOMS 2018-2018 IEEE/IFIP Network Operations and Management Symposium*, pages 1–6. IEEE, 2018.

[36] Sunday O Oladejo and Olabisi E Falowo. 5g network slicing: A multi-tenancy scenario. In *2017 Global Wireless Summit (GWS)*, pages 88–92. IEEE, 2017.

[37] Salvatore Costanzo, Ilhem Fajjari, Nadjib Aitsaadi, and Rami Langar. A network slicing prototype for a flexible cloud radio access network. In *2018 15th IEEE Annual Consumer Communications & Networking Conference (CCNC)*, pages 1–4. IEEE, 2018.

[38] Claudia Campolo, Antonella Molinaro, Antonio Iera, and Francesco Menichella. 5g network slicing for vehicle-to-everything services. *IEEE Wireless Communications*, 24(6):38–45, 2017.

[39] Ramon Dos Reis Fontes, Claudia Campolo, Christian Esteve Rothenberg, and Antonella Molinaro. From theory to experimental evaluation: Resource management in software-defined vehicular networks. *IEEE access*, 5:3069–3076, 2017.

[40] Number of connected devices worldwide in 2014 and 2020, by device. https://www.statista.com/statistics/512650/worldwide-connected-devices-amount. Accessed: 2022-05-27.

[41] Ruben Solozabal, Aitor Sanchoyerto, Eneko Atxutegi, Bego Blanco, Jose Oscar Fajardo, and Fidel Liberal. Exploitation of mobile edge computing in 5g distributed mission-critical push-to-talk service deployment. *IEEE Access*, 6:37665–37675, 2018.

[42] Tuyen X Tran, Abolfazl Hajisami, Parul Pandey, and Dario Pompili. Collaborative mobile edge computing in 5g networks: New paradigms, scenarios, and challenges. *IEEE Communications Magazine*, 55(4):54–61, 2017.

[43] Bego Blanco, Jose Oscar Fajardo, Ioannis Giannoulakis, Emmanouil Kafetzakis, Shuping Peng, Jordi Pérez-Romero, Irena Trajkovska, Pouria S Khodashenas, Leonardo Goratti, Michele Paolino, et al. Technology pillars in the architecture of future 5g mobile networks: Nfv, mec and sdn. *Computer Standards & Interfaces*, 54:216–228, 2017.

[44] Yun Chao Hu, Milan Patel, Dario Sabella, Nurit Sprecher, and Valerie Young. Mobile edge computing – a key technology towards 5g. *ETSI white paper*, 11(11):1–16, 2015.

[45] Broadband commission deliberates building an inclusive post-covid digital future. https://www.itu.int/en/mediacentre/Pages/pr03-2021-Broadband-Commission-inclusive-post-COVID-digital-future.aspx. Accessed: 2022-05-27.

[46] Mary-Ann Russon and Lucy Hooker. Uk 'heading towards digital skills shortage disaster'. https://www.bbc.co.uk/news/business-56479304. Accessed: 2022-05-27.

[47] Digital education action plan (2021-2027). https://education.ec.europa.eu/focus-topics/digital-education/about/digital-education-action-plan. Accessed: 2022-05-27.

[48] Elearning market to exceed usd 375 billion in 5 years. https://learningnews.com/news/learning-news/2020/elearning-market-to-exceed-usd-375-billion-in-5-years-(1). Accessed: 2022-05-27.

[49] Nasir Saeed, Ahmed Bader, Tareq Y Al-Naffouri, and Mohamed-Slim Alouini. When wireless communication responds to covid-19: Combating

the pandemic and saving the economy. *Frontiers in Communications and Networks*, 1:566853, 2020.

[50] Arghir-Nicolae Moldovan and Cristina Hava Muntean. Dqamlearn: Device and qoe-aware adaptive multimedia mobile learning framework. *IEEE Transactions on Broadcasting*, 67(1):185–200, 2020.

[51] Longhao Zou, Ramona Trestian, and Gabriel-Miro Muntean. E 3 doas: Balancing qoe and energy-saving for multi-device adaptation in future mobile wireless video delivery. *IEEE Transactions on Broadcasting*, 64(1):26–40, 2017.

[52] Ramona Trestian, Ioan-Sorin Comsa, and Mehmet Fatih Tuysuz. Seamless multimedia delivery within a heterogeneous wireless networks environment: Are we there yet? *IEEE Communications Surveys & Tutorials*, 20(2):945–977, 2018.

[53] Ting Bi, Longhao Zou, Muhammed Maddi, Gregor Rozinaj, and Gabriel-Miro Muntean. Knowledge acquisition by employing adaptive multimedia in third level technology enhanced learning stem education. In *EdMedia+ Innovate Learning*, pages 1773–1779. Association for the Advancement of Computing in Education (AACE), 2019.

[54] G-M Muntean. Efficient delivery of multimedia streams over broadband networks using qoas. *IEEE Transactions on Broadcasting*, 52(2):230–235, 2006.

[55] Longhao Zou, Ting Bi, and Gabriel-Miro Muntean. A dash-based adaptive multiple sensorial content delivery solution for improved user quality of experience. *IEEE Access*, 7:89172–89187, 2019.

[56] Zhenhui Yuan, Gheorghita Ghinea, and Gabriel-Miro Muntean. Beyond multimedia adaptation: Quality of experience-aware multi-sensorial media delivery. *IEEE Transactions on Multimedia*, 17(1):104–117, 2014.

[57] Abid Yaqoob, Ting Bi, and Gabriel-Miro Muntean. A survey on adaptive 360 video streaming: solutions, challenges and opportunities. *IEEE Communications Surveys & Tutorials*, 22(4):2801–2838, 2020.

[58] Nabajeet Barman and Maria G Martini. Qoe modeling for http adaptive video streaming–a survey and open challenges. *IEEE Access*, 7:30831–30859, 2019.

[59] Adam Polakovič, Radoslav Vargic, and Gregor Rozinaj. Adaptive multimedia content delivery in 5g networks using dash and saliency information. In *2018 25th International Conference on Systems, Signals and Image Processing (IWSSIP)*, pages 1–5. IEEE, 2018.

[60] Ioan-Sorin Comşa, Andreea Molnar, Irina Tal, Per Bergamin, Gabriel-Miro Muntean, Cristina Hava Muntean, and Ramona Trestian. A machine learning resource allocation solution to improve video quality in remote education. *IEEE Transactions on Broadcasting*, 67(3):664–684, 2021.

[61] Irene-Angelica Chounta, Emanuele Bardone, Aet Raudsep, and Margus Pedaste. Exploring Teachersâe^{TM} Perceptions of Artificial Intelligence as a Tool to Support their Practice in Estonian K-12 Education. *International Journal of Artificial Intelligence in Education*, June 2021.

[62] Xusen Cheng, Jianshan Sun, and Alex Zarifis. Artificial intelligence and deep learning in educational technology research and practice, 2020.

[63] Mutlu Cukurova, Carmel Kent, and Rosemary Luckin. Artificial intelligence and multimodal data in the service of human decision-making: A case study in debate tutoring. *British Journal of Educational Technology*, 50(6):3032–3046, 2019.

[64] Cristina Conati, Kaska Porayska-Pomsta, and Manolis Mavrikis. Ai in education needs interpretable machine learning: Lessons from open learner modelling. *arXiv preprint arXiv:1807.00154*, 2018.

[65] Mary E Webb, Andrew Fluck, Johannes Magenheim, Joyce Malyn-Smith, Juliet Waters, Michelle Deschênes, and Jason Zagami. Machine learning for human learners: opportunities, issues, tensions and threats. *Educational Technology Research and Development*, 69(4):2109–2130, 2021.

[66] Almaas A Ali, Georgios A Dafoulas, and Juan Carlos Augusto. Collaborative educational environments incorporating mixed reality technologies: A systematic mapping study. *IEEE Transactions on Learning Technologies*, 12(3):321–332, 2019.

[67] George Dafoulas, Noha Saleeb, and Martin J Loomes. What learners want from educational spaces? a framework for assessing impact of architectural decisions in virtual worlds. *IADIS International Journal on WWW/Internet*, 14(2):72–90, 2016.

[68] Martyn Wyres and Natasha Taylor. Covid-19: using simulation and technology-enhanced learning to negotiate and adapt to the ongoing challenges in uk healthcare education. *BMJ Simulation & Technology Enhanced Learning*, 6(6):317–319, 2020.

[69] Xiantao Jiang, F Richard Yu, Tian Song, and Victor CM Leung. A survey on multi-access edge computing applied to video streaming: some research issues and challenges. *IEEE Communications Surveys & Tutorials*, 23(2):871–903, 2021.

[70] Gabriel-Miro Muntean and Nikki Cranley. Resource efficient quality-oriented wireless broadcasting of adaptive multimedia content. *IEEE Transactions on Broadcasting*, 53(1):362–368, 2007.

[71] Altair. Altair winprop 2021.1 user guide. *Altair Engineering Inc.*, 2021.

Part III

Paradigms of Intelligent Networked Systems

Chapter 7

Intelligent Environments and Internet of Things

"We will always have STEM with us. Some things will drop out of the public eye and will go away, but there will always be science, engineering, and technology. And there will always, always be mathematics."

Katherine Johnson
Mathematician at NASA,
Awarded the Presidential Medal of Freedom,
America's highest civilian honour, by President Obama.

7.1 IoT Life-Cycle

Internet of Things (IoT) is the network of physical objects that are referred to as *things*. These objects are integrated with particular sensors, software protocols and network adapters to enable intra/inter-connection of objects. With their unique identifiers, these objects are able to associate to each other and collect, transfer and exchange data over the Internet without requiring any human intervention. Therefore, IoT enables remote sensing and controlling of devices/objects over the Internet through existing network infrastructure. IoT is considered to be the next step in the Internet evolution as it opens up new opportunities for players in various industries, such as embedded and control systems, home automation, healthcare, automotive engineering, consumer electronics, education, manufacturing, etc. In this context, the rapid growth of smart devices, sensors, wireless networks, big data and computing power will accelerate the development and deployment of massive IoT. However, in practice, there are substantial issues that need to be addressed with IoT, such as optimization of energy consumption and carbon emission, handling network scaling and complexity, determining device proximity, peer-to-peer connections, low-latency for real-time interaction, integration of devices that have little or no processing capability, etc. [1].

Figure 7.1 presents the IoT life-cycle [2]. The first component are the *sensors* that are represented by small hardware devices that detect events or

DOI: 10.1201/9781003222095-7

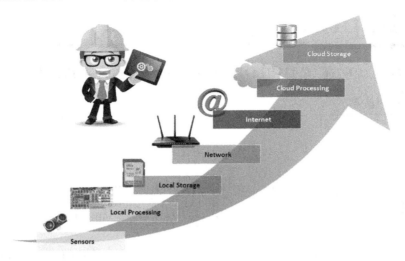

FIGURE 7.1
IoT Life-Cycle

changes in the environment and produce an output signal like a measurement as a response to a physical stimulus like heat, light, sound, pressure, etc. In General, the sensor devices have an analog input (e.g., temperature, wind speed, etc. measurement) and convert it to digital data/electronic data. Most of the sensors do not do any processing or very little processing. For example, the processing required to convert the data from the environment into something electronic, so analog to digital conversion. They usually consume low power because they need to run for a long period of time depending on the applications. These sensor devices can be mounted outside, on a bike, on a car, etc. representing the *things* in the IoT that collect data and send the information to other electronics or some kind of processing device like a computer processor. This takes us to the next step in the IoT life-cycle illustrated in Figure 7.1, which is the *local processing*. This is represented by some computing device that collects the data from the sensors and performs some local processing in order to take some decisions. The next step in the IoT life-cycle is the *local storage* as some of the data collected from the sensors requires to be stored locally. The next step is the *network* to be able to interconnect the *things*. In order to enable the Internet in the IoT and we also need the Internet as the next step to be able to send the data to a *cloud processing* platform that has more resources then the local processing, thus it can perform a more computational intensive processing on the sensor data. The final step in the IoT life-cycle is the *cloud storage* to store the data.

This represents the IoT life-cycle and in fact it represents an ideal scenario. This is because in the current environment the *local processing* and *local storage* are skipped and all the data from the sensors goes straight to the cloud for computationally intense processing. Consequently, a significant amount of data is collected from the sensors and sent to the cloud which needs to process it. However, most of the data might not be relevant and would not have required to be sent to the cloud. An example would be the scenario of a motion sensor within an office environment, that collects the data periodically (e.g., every 1 second) and sends it to the cloud for processing. However, not all this data is relevant as there might not be any activity outside the office hours. Consequently, within an ideal IoT life-cycle the data from the motion sensor would be sent to a local processing to identify if there are any anomalies. For example, an intruder in the office room that would trigger the data to be sent to the cloud and alert the police. Thus, the data that is already processed by the local processing device should be send to the cloud. The cloud should aggregate this data, compute results and offer some predictions as well as store the data on the long term [2].

The local processing or local storage could be done by Edge (Intel) or Fog (Cisco) computing because the computation is moved closer to the user. As we have seen, some processing might be more convenient to happen locally. However, there is a trade-off that needs to be achieved depending on the application, between the energy consumption and the local processing.

The first step into democratizing the electronics was represented by the open hardware movement. The idea behind *open hardware* or *open source hardware* is that all the information regarding the hardware specifications is freely available to the public. This concept gave a boost into the advancements of IoT as it enabled more things to be connected to the Internet.

Some of the most widely used IoT development boards are illustrated in Figure 7.2 The first example is the Arduino board based on the ATmega328 chip that is a small microcontroller and the most used one. There are several flavours of Arduino available on the market with Arduino UNO the most common. ChipKIT is build by Digilent on top of different microcontroller referred to as Programmable Integrated Circuit (PIC) which is a little bit more powerful than Arduino. LaunchPad was build by Texas Instruments using the MSP430 chip, consumes less power than the rest and is mainly used to simulate hardware electronic components like Registers, Gates, etc. These devices are good for connecting sensors. Some hardware devices that have improved processing are the STM32 chips which come with ARM Cortex M0, M3, M4 depending on how much processing power is required. Particle has WiFi integrated. While Espruino stands out from the rest due to the fact that is using the Javascript as compared to most of the other microcontrollers that are using C or some other languages that compile mostly C [2].

Raspberry Pi is also widely used, as it has 900 MHz dual core chip and 1 GB RAM storage on SD card. The Beaglebone Black appeared before Raspberry Pi but it was more expensive. It has 1 GHz ARM CPU, GPU integrated, 512 MB RAM and it also has a 4 GB flash storage but an SD card can be accommodated. UDOO Neo represents a combination of Raspberry Pi and Arduino and it has a free scale microprocessor ARM for microcontroller. There

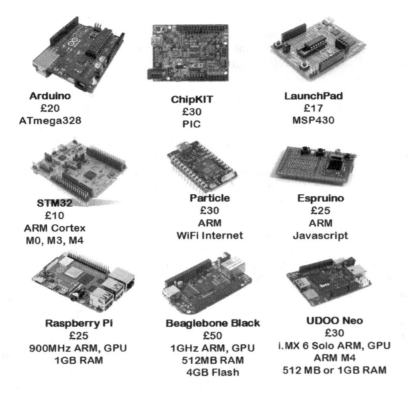

FIGURE 7.2
Examples of Hardware for IoT

are many other hardware devices for IoT development. However, these are the most common at the moment.

7.2 IoT Applications

Nowadays we are living in a connected world, where billions of devices and increasingly every thing is connected to the Internet. All these devices and things are capable of constantly collecting, transmitting and analyzing any kind of data.

The wireless connectivity enables the realization of IoT and boosts innovations. The benefits of IoT could be considered and taken advantage of in order to boost different areas within the industry, with the greatest impact expected to be on: autonomous vehicles/connected transportation, smart cities, smart

FIGURE 7.3
Connected World

agriculture, smart healthcare, smart entertainment, etc. as depicted in Figure 7.3.

Apart from IoT, another game changer is the large scale adoption of video-based applications and social networks. It is predicted that 75% of the 5G network spectrum will be used to transmit video signals [3]. However, by integrating multimedia communication and IoT there could be a dramatic change in the way people are living. For example, in case of disaster management the first responders could get real-time multimedia streaming data of the incident before they actually reach the scene. In this way they could anticipate the scale and emergency of the disaster. Other areas that rich media IoT could have a significant impact on are: healthcare, education, intelligent transportation, agriculture, etc.

However, it is known that multimedia traffic is power and bandwidth-hungry and has strict Quality-of-Service (QoS) requirements. Thus, the technical challenges posed by transmitting real-time rich media data over wired and/or wireless links within IoT need to be identified and addressed.

The future rich media IoT-5G integration could open up opportunities for various intelligent solutions, such as: smart video-based security systems, intelligent unmanned aerial vehicles (UAV)/drones, remote robotic-controlled surgery, smart homes, smart parking, smart farming, etc.

The major shift in paradigm will happen when the surrounding objects and things will start being connected on a social level as well, creating in this way the Internet of Social Things (IoST). However, to accommodate this IoST, the IoT objects need to have a social life that enable them to collaborate around a shared purpose and become social.

A good example of IoST is represented by Waze[1]. Waze is a real-time community-based traffic and navigation application which enables users to update traffic-related information, such as: road conditions, traffic intensity, accidents, etc. This information is then used to find the best route to get to a destination. This represents a good example of how the connected cars could take advantage of the collective intelligence of the community of connected cars.

IoST is envisioned to be applied to other areas as well, such as connecting medical devices and people by forming a social network of doctors, patients and family members [4]. Or the connected shoes provided by Nike [5] so that the customers could compete against each other in various sport activities.

7.2.1 Smart Surveillance Systems

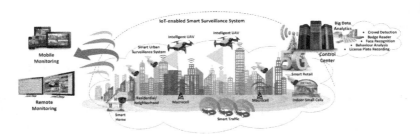

FIGURE 7.4
Smart Surveillance Example

One of the most common and important rich media-based IoT application is represented by the smart surveillance systems which are mainly used for real-time road traffic monitoring, security control, smart parking, etc. These smart surveillance systems, consist of cameras or imaging sensors installed in key locations depending on a specific purpose, such as smart transportation and urban surveillance applications, smart homes, smart retail, etc. as indicated in Figure 7.4. The data from the surveillance systems is sent to a Control Center (e.g., cloud) where is stored and analyzed. Depending on the application requirements, the data at the Control Center is processed using some key techniques, such as: crowd detection, badge reader, face recognition, behavior analysis, license plate recording, etc. In general, these techniques require high-complexity image/data processing technologies and algorithms that need to provide accurate and reliable results. For example, in case of license plate recognition several methods have been adopted in the literature, such as: image processing [6], profile-based filter [7], contrast between grayscale values [8], morphology-based solutions [9; 10] color-based methods [11], unified

---

[1]https://www.waze.com/en-GB/

method of automated object detection [12] or more recently, convolutional neural networks [13; 14].

In the context of smart home [15] a smart surveillance system will monitor the house and evaluate through data processing is an intruder is detected of not. Upon the detection of an intruder, the house alarm will be activated as well as a notification will be sent to the user's mobile device.

The smart surveillance systems could also enable the implementation of highly accurate video-based indoor positioning systems [16]. These systems could be used for improving the security in smart retail by providing real-time and accurate indoor positioning of the customers.

These smart surveillance systems represent key enablers for *Smart Cities* aiming at improving the quality of live for the citizens. One interesting use case scenario is described in [17], where the authors tackle the issues of safety at the beach side. In this context, smart surveillance cameras have been deployed and the video-based surveillance is used for downing prevention by detecting and monitoring the human activity in the water.

Additionally, the integration of intelligent unmanned aerial vehicles (UAV) within the smart surveillance systems has been proposed in [18]. In this context, the police could deploy patrol UAVs to act as mobile surveillance platforms where the fixed CCTV cameras are not in place or the target is on the move. This solution could increase the security and reduce crime related costs.

However, the integration of these solutions within the concept of *Smart Cities* will require a number of IoT devices greater than 1000 million per city, generating a massive amount of video-based data traffic at an approximate capacity of more than 10-100 million GB of real-time data per city per day with an expected latency of less than 1ms [19]. This represents a great challenge even for the highly anticipated 5G networks.

7.2.2   Autonomous Vehicles

One exciting vision for the next generation networks is the autonomous vehicles, where most of the cars on the road will be connected and controlled through IoT. This will facilitate information collection, processing and analysis that could be used to improve transportation efficiency, ensure traffic safety and reduce traffic load. Within industry, there are currently a number of companies already working on driverless cars, such as: Tesla, Uber, Toyota, Hyundai, Waymo, BMW, Ford, etc.

As illustrated in Figure 7.5, the IoT data collected from cars, pedestrians, road side infrastructure, etc. could be sent to a Control Center for storage, processing and analysis. However, to enable fully automated vehicles, a highly accurate and performing intelligent control system is required to detect and avoid obstacles, pedestrians, cyclists and other road users. This is achieved by the wide range of IoT objects and sensor technologies deployed as well as the V2X communication.

One of the core technologies for self-driving vehicles is the use of High Definition (HD) maps [20]. The HD maps give highly accurate precision at

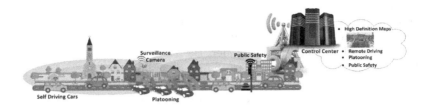

FIGURE 7.5
Autonomous Vehicles Example

centimeter level which is very important especially for driving in crowded urban areas. The HD maps are created by combining the information received from the IoT objects and LiDAR (Light Detection and Raging). LiDAR works by sending and tracking down light pulses in order to identify the surrounding objects. The benefits of using LiDAR is that it overcomes the drawbacks of Radar (radio waves) for short range and Sonar (sound waves) for long range. Consequently, LiDAR can accurately identify and map the near and far objects/obstacles on the road. Additionally, in order to improve the efficiency and quality of the HD Maps, machine learning could be integrated [21] to create a self-learning platform that continuously analyzes the data on the fly and introduces safety improvements.

To enable autonomous driving, the number of IoT devices required per vehicle is somewhere around 50-200 objects generating a massive amount of real-time data including rich media with a capacity of more than 100 GB per vehicle per day and with an expected latency of less than 1ms [19].

7.2.3 Healthcare

The penetration of IoT-based solution into healthcare led to the appearance of Internet of Health Things (IoHT) [22]. IoHT enables the connectivity between the patient and the healthcare facilities. The use of biomedial sensors help the individual to monitor any of the vital signal of the body, including heart rate, blood pressure, body temperature, oxygen in blood, airflow, etc. As illustrated in Figure 7.6, the data from these IoT objects could be collected, stored and analyzed in the cloud [23]. The data could be accessed by the doctor either in the Hospital or remotely in real time. Additionally, the health data analysis could provide suggestions for health improvements to the patients through their mobile devices.

Real time video streaming for remote patient monitoring and real-time statistics observation could be combined with the data from other IoT objects to help doctors in providing more accurate diagnosis and advice for their

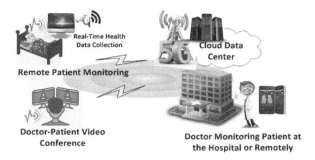

FIGURE 7.6
Smart Healthcare Example

patients. Moreover, there is also the possibility to enable real-time video conference between doctors and patients regardless of their location. In this way, the hospital costs and clinical time are reduced while the quality of care is increased.

The integration of IoT within ambient assisted living has seen great potential in offering support especially for elderly people enabling them to live a more independent but safe lifestyle [24]. By using wearable and ambient IoT devices and sensors, one could monitor the daily living activities of the ageing population, their social interactions and take control over their chronic disease management in order to improve their wellbeing.

The continuous advances in technology led to the development of remotely controlled robotic surgeries or computer-assisted surgeries where machines carry out surgical procedures. In case of remote surgery, the number of IoT devices required is somewhere more than 10-100 per surgery, generating real-time streaming data with an expected latency of less than 200ms [19].

7.2.4 Virtual Reality

The combination of IoT and Virtual Reality (VR) led to a new paradigm referred to as Virtual Environment of Things (VEoT) [25; 26]. The concept of VEoT enables the real-time interaction between avatars/objects from the computer generated virtual environment and real-world smart IoT objects. VEoT has the potential to revolutionize the entertainment world, especially the gaming industry as illustrated in Figure 7.7. By using smart IoT objects and VR equipment users can now join the game world and interact with other real-world VR users or computer-generated virtual avatars and objects.

The combined use of VR and Augmented Reality (AR) with the urban IoT data streams from different spatial locations was explored in the context of creating a Mixed Reality (MR) environment where the digital data visualizations and the physical space are mapped so that the citizens could collaborate, interact and navigate naturally [27].

FIGURE 7.7
Virtual Reality Example

VR and IoT have also been integrated in order to maximize the high interoperability of IoT services with the intuitive nature of VR [28; 29]. This integration also allowed the creation of an immersive virtual representation of a smart city with remote sensing [30]. These VR-IoT integrated platforms can provide a customizable and user-friendly environment for controlling real IoT devices in virtual environments.

The combination of AR and IoT sensors is also used to connect the virtual game content to the real world as a way to motivate people to exercise and overcome obesity [31]. These games are referred to as exergames that could potentially improve youths' health status as well as provide social and academic benefits [32].

The number of VR and AR IoT devices is more than 0.2 million globally generating a massive amount of data streaming traffic at data rates of more than 1Gbps and with an expected latency of less than 1ms [19].

7.2.5 eLearning

Until recently, the concept of smart classroom/school referred to the classrooms that had integrated one type of smart device, like the smart board. However, nowadays the concept of smart school makes use of IoT as a computational system that collects student-related data from the surrounding smart devices (e.g., eBooks, tablets, various sensors, smart displays, etc.) to provide a personalized student learning experience as well as enabling remote or on the move eLearning as depicted in Figure 7.8. By creating this multimedia-based IoT-centric student learning environment enables flexibility for teachers and students enabling them to access the learning content from anywhere at anytime and from any device [33].

Moreover, the educational system could avail of IoT together with AR technology to provide a more relevant, engaging and interactive courseware to the students to keep them motivated and interested in the subject [34]. AR gaming enables the delivery of the learning content through mobile application

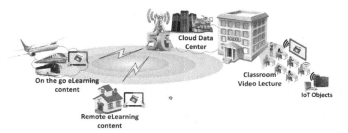

FIGURE 7.8
eLearning Example

games that facilitate the learning as well as the knowledge evaluation through playing.

7.2.6 Smart Agriculture

The use of IoT in Agriculture has seen great potential especially for precision agriculture representing an useful asset for farmers. In this context, the IoT devices could provide precise information about the production and different environmental factors: soil quality, irrigation levels, pest and fertilizer [35] or green house production environment [36]. To enable the smart agriculture environment the IoT sensors could be deployed on tractors, in the field and soil or intelligent UAVs could be used to collect video-based information from above as illustrated in Figure 7.9. All this information collected from the IoT sensors could be sent to a central cloud for storage, processing and analysis.

Various information management systems approaches could be used to aid farmers in analysing and interpreting the data in order to optimize the growing environment and promote smart cultivation at any point within the production chain [37]. Consequently, integrating IoT within agriculture could

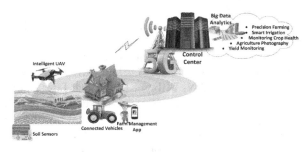

FIGURE 7.9
Smart Agriculture Example

enhance productivity and also enable opportunities for automating farming methods.

7.3 Wireless Access Networks for IoT

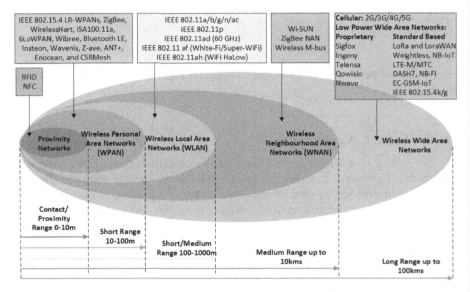

FIGURE 7.10
Wireless Access for IoT Based on Geographic Coverage

Figure 7.10 illustrates a comparison and classification in terms of coverage area of various wireless access networks for IoT applications. The shortest transmission range of up to 10m is given by the close proximity networks that are based on Radio Frequency Identification (RFID) and Near Field Communication (NFC). The next category of networks is the Wireless Personal Area Network (WPAN) that can reach up to 100m coverage area and include technologies like IEEE 80.15.4 low rate WPANs, ZigBee, WirelessHart, ISA100.11a, 6LoWPAN, Wibree, Bluetooth Low Energy (BLE), Insteon, Wavenis, Z-wave, ANT+, Enocean, and CSRMesh. We can notice that most of these technologies are designed for low data rate, low power and short distance. Increasing the transmission range further to up to 1000m we have the WLANs that are designed for higher data rates and include different standards of the IEEE 802.11 family. The next category is the wireless Neighborhood Area Network (WNAN) with a transmission range of up to 10 kms can be used for residential, campus, street-level environments and for utility and smart grid applications. Thus, Wireless Meter-bus known as smart metering belongs to this category. The category that can achieve the highest transmission range is the Wireless Wide Area Network (WWAN) that include both cellular networks and Low Power Wide Area Network (LPWAN). The LPWAN further consists of proprietary and standard-based solutions [38]. Due to the diversity of applications these wireless access solutions are there to complement each other and not to compete against each other. Furthermore, some of the future intelligent applications might require the use of a combination of these wireless access solutions.

Table 7.1 illustrates the mapping of the wireless technologies to the key IoT verticals. Due to the fact that each IoT vertical and application has its own set of network requirements, we can see that each IoT wireless technology is relevant for different IoT Verticals.

TABLE 7.1
IoT Wireless Technologies vs. IoT Verticals

Key IoT Verticals	LPWAN (Star)	Cellular (Star)	ZigBee (Mostly Mesh)	BLE (Star & Mesh)	WiFi (Star & Mesh)	RFID (Point-to-Point)
Industrial IoT	√√	√	√			
Smart Meter	√√					
Smart City	√√					
Smart Building	√√		√	√		
Smart Home			√√	√√	√√	
Wearable	√			√√		
Connected Car					√	
Connected Health		√√		√√		
Smart Retail		√		√√	√	√√
Logistics & Asset Tracking	√	√√				√√
Smart Agriculture	√√					

√√- Highly applicable;
√- Moderately applicable;

7.3.1 LoRa and LoRaWAN

As we have seen, there are many wireless technologies that can be used to connect the *things* in the IoT to enable them to communicate. In general the LPWAN have the advantage of long range and low power characteristics which makes them suitable for the IoT applications where remote locations, easy deployments, high number of devices, and long battery life are required. The LPWAN are designed for small data packages, thus small data rates that need to be send over long distances between devices that are operating on battery.

One LPWAN technology that can accommodate these requirements and which belongs to the WWAN class is Long Range (LoRa). Using LoRa, the battery powered or low powered wireless nodes can transmit data at low data rates (e.g., 0.3kbps to 5.5kbps) wirelessly over long distances. However, the distance between a transmitter and a receiver also depends on the device characteristics as well as the environment it operates in. For example, in a direct LoS communication the communication distance can reach more than 15 km.

Figure 7.11 illustrates the LoRa protocol stack [39] mapped to the OSI reference model. It can be noted that LoRa corresponds to the PHY layer while LoRaWAN is built on top of LoRa and corresponds to the MAC and Application layer for the communication network

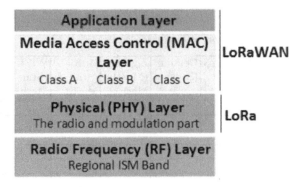

FIGURE 7.11
LoRa Protocol Stack

There are two types of LoRa devices: (1) the *LoRa end node* which is a battery powered device that consists of the radio module and an antenna as well as a microprocessor for some local processing of the sensor data. If the sensors are located on the device then we can refer to the device as a *mote*. (2) the *LoRa gateway* which is usually mains powered and also consists of the radio module and an antenna as well as a microprocessor for processing of the data. One LoRa end node device can send data to multiple gateways and the communication is bidirectional, meaning that the end node can also receive data from the gateway.

The LoRa PHY defines the radio and the type of modulation used. LoRa is using a proprietary spread spectrum modulation scheme that is based on the Chirp Spread Spectrum (CSS) modulation that makes use of wide-band linear frequency modulated *chirp* pulses for encoding the information required to be transmitted. The idea behind the spread spectrum is to spread the narrow-band information signal over a wide-band in order to make the communication more secure and more difficult to intercept. In the case of CSS, there are two types of *chirp* or *sweep signals*, such as: *up-chirp* corresponding to an

TABLE 7.2

LoRa Spreading Factors for 125 kHz Bandwidth

Spreading Factor	Chips/ symbol	SNR limit	Time-on-air (10 byte packet)	Bitrate
7	128	−7.5	56 ms	5469 bps
8	256	−10	103 ms	3125 bps
9	512	−12.5	205 ms	1758 bps
10	1024	−15	371 ms	977 bps
11	2048	−17.5	741 ms	537 bps
12	4096	−20	1438 ms	293 bps

increase in frequency with time and *down-chirp* corresponding to a decrease in frequency with time. For example, in the case of down-chirp, it starts with the high frequency which decreases in time until it reaches the low frequency when it goes back to the high frequency and it repeats the process.

LoRa makes use of the unlicensed sub-GHz radio frequency bands, that might vary from country to country. For example, in Europe LoRa makes use of the 868 MHz band which enables a larger coverage area due to easier penetration through objects at lower frequencies. Moreover, LoRa employs an adaptive data rate by using a combination of bandwidth and the spreading factor. The spreading factor indicates the number of raw bits that can be encoded by the symbol. Depending on the quality of the communication link and the amount of data to be transmitted a certain bandwidth and spreading factor is selected. For example, a higher spreading factor is used when the end node is locate further away from the gateway, which means the signal quality is weak. Table 7.2 [40] indicates the LoRa spreading factors for 125 kHz bandwidth. It can be noted that a higher data rate is achieved for a low spreading factor but less transmission range, while for for larger transmission range the spreading factor is higher, the data rate is lower and the time-on-air also increases. Consequently, there is a trade-off between the transmission range and the data rate.

As we have seen, LoRa defines the PHY layer, and while there are many other proprietary layers that can be built on top of it, the LoRaWAN is the most popular MAC layer. The LoRaWAN network architecture is illustrated in Figure 7.12 [41] and consists of four major components, such as: the end nodes/devices, the gateways, the network server, and the application server. The topology uses in LoRaWAN is a star network topology where the end nodes communicate directly with the gateway, since they have enough transmission range. This makes the entire network simple and easy to manage.

The end nodes collect data from the environment they are located in, this data is then broadcast to the neighbouring gateways which will forward it to the network server. The network server aggregates all the data from the gateways and removes the duplicates before forwarding it to the specific application server. As the communication can be bidirectional if required, the

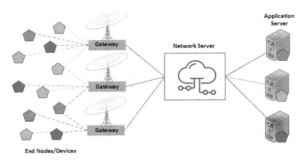

FIGURE 7.12
LoRaWAN Network Architecture

application server can respond back to the network server, which selects the gateway with the best reception to broadcast the response back to the end node.

There are three classes of devices defined by LoRaWAN, such as:

- *Class A*: represented by battery powered sensors. The devices in this class require donwlink communication from the server only after the sensor transmission which enables them to keep the communication at the minimum and thus increase the battery lifetime. However, increase the communication latency. This is the most energy efficient class.

- *Class B*: represented by battery powered actuators. The devices in this class require extra downlink windows at scheduled intervals apart from the downlink communication link after the transmission. Thus it keeps the same characteristics of Class A plus a scheduled extra downlink. By using a scheduled approach it can still control power consumption. However, this class of devices have a lower battery life as compared to the class A devices but it brings improvements in terms of latency.

- *Class C*: represented by the main powered actuators. The devices in this class can communicate continuously bringing the communication latency to a minimum. However, this device class has the lowest battery life as compared to the other two.

The main advantages of LoRa and LoRaWAN are the long range and coverage that cannot be compared with that of any other technology, the low power radios to accommodate at least 10 years of usage on a single battery charge, the low cost hardware and the availability of open source infrastructure devices, as well as the high capacity with thousands of end node devices being able to connect to one LoRa gateway. The only disadvantage is the low data rate that might not be ideal for specific applications that require higher data rates.

7.3.2 ZigBee

ZigBee is a wireless technology that extends the IEEE 802.15.4 standard by building four main components on top, such as: the network layer to provide routing; an application support sublayer to enable the use of specialized services; ZigBee device objects used to maintain device roles, manage network join requests, device discovery and manage security; and manufacturer-defined application objects that enable customization.

ZigBee was designed for monitoring and controlling sensor networks and operates in the 868 MHz band in Europe, 915 MHz in USA and Australia and 2.4 GHz worldwide. The aim of ZigBee was to enable a very low cost implementation of low power devices with low data rates of between 20 to 250 kbps over a transmission distance of 300+ meters LoS. A ZigBee network can scale up to 240 connected devices. There are three types of ZigBee devices, such as: (1) *coordinator* – which is the root of the network being the most capable device responsible for several tasks: channel selection, network ID allocation, device unique address allocation, creates, controls and maintains the network. Each ZigBee network must have one coordinator. (2) *routers* – are responsible of routing the traffic between different nodes and the coordinator, they receive and store messages intended for the end devices connected to them, they can allow other routers and end devices to join the network; (3) *end devices* - represented by the battery operated devices (e.g., sensors) that can be in the sleep or standby mode most of the time to conserve the battery. They communicate to the router node they are connected to and they are responsible for requesting any pending information from them that was transmitted while in sleep mode. They cannot relay information.

Figure 7.13 illustrates the possible topologies of a ZigBee network and described below:

- *star topology* – this is the simplest and less expensive implementation that consists only of one coordinator device and multiple end devices that communicate with the coordinator. However, if the coordinator device fails the entire network fails. Additionally, the network range is limited to the range of the coordinator.

- *mesh topology* – in this case every router device is connected with its neighboring router device and the messages hop from one router to another in order to reach the destination. In case a node in the network fails, data can be reroute on an alternate path.

- *tree topology* – similar to mesh configuration except the routers are not interconnected.

- *cluster tree topology* – aims to expand the network range.

The coordinator is responsible with channel allocation. There will be only one channel assigned to the network and two channel access methods are

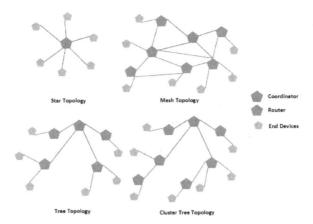

FIGURE 7.13
ZigBee Network Topologies

defined: (1) the contention-based method is CSMA/CA so that all the devices can share the single channel and enable communication without being synchronized and (2) the contention free method where the coordinator will transmit periodically the time slot configuration for each end device referred to as guaranteed time slot. Additionally a beacon is used to synchronize the clocks of all the devices in the network.

The most prominent applications that can benefit from ZigBee are in home automation, medical data for remote monitoring system, industrial control, etc. The main purpose is to use ZigBee to collect information or perform certain control tasks inside buildings.

7.3.3 IEEE 802.11ah

The IEEE 802.11ah wireless technology also known as Wi-Fi HaLow, offers longer range and lower power connectivity as compared to the traditional Wi-Fi networks. IEEE 802.11ah operates in the sub-1 GHz license exempt spectrum within the 900 MHz band (varies according to the country). The sub-1 GHz frequency offers longer range than 2.4 GHz, due to better penetration of the signal through obstacles and it is less congested. The IEEE 802.11ah standard was developed for use in smart building applications, like smart lighting, smart heating, ventilation, and air conditioning (HVAC), smart security systems, connected cars, smart cities and other low power applications. The standard defines new PHY and MAC layers with the PHY layer similar to a sub-1 GHz version of the IEEE 802.11ac standard. Channel bandwidths can vary from 1, 2, 4, 8 and 16 MHz, depending on the specific country. PHY transmission is based on OFDM and the supported modulations include BPSK, QPSK and 16, 64 and 256-QAM.

Moreover, the IEEE 802.11ah standard also adopts technologies like single-user beamforming, MIMO, MU-MIMO for downlink, etc. which were first introduced in the IEEE 802.11ac.

In order to be able to support a large number of devices per AP as required by IoT, the MAC layer introduces a novel hierarchical method that defines different groups of stations [42] by using a hierarchical Association Identifier (AID). The AID consists of four hierarchical levels defined as: pages, blocks, sub-blocks, and stations. Consequently, eight stations are indexed in a given subblock, eight subblocks are indexed in a specific block, and 32 blocks belong to a certain page. This means that a page can contain up to 2048 stations. However, the AID structure includes four pages, which means that the AP can support up to 8191 stations. This represent a significant increase in the network size for a Wi-Fi network.

The IEEE 802.11ah standard also defines three types of stations that require different procedures and time periods to access the transmission channel: (1) *Traffic Indication Map (TIM) stations* – represented by stations with a high traffic load that must perform the data transmission within a restricted access window (RAW); (2) *non-TIM stations* – represented by station with periodic low traffic that directly negotiate a transmission time with the AP allocated in the periodic RAW (PRAW); (3) *unscheduled stations* – represented by stations with very low traffic that can send a poll frame to the AP to request immediate access to the channel.

7.3.4 Bluetooth

Bluetooth is a single-chip, low cost radio-based wireless network technology that is mainly used for connecting different small devices and/or peripherals in an wireless ad-hoc manner without any infrastructure required. Similarly to other technologies, Bluetooth is using the license-free 2.4 GHz band. The typical range of Bluetooth is around 10 meters making it a WPAN technology.

The basic network topology of Bluetooth is called *piconet* consisting of one master device and multiple slave devices. While the overlapping piconets (stars) form a scatternet as illustrated in Figure 7.14.

There can be only one master in a piconet and up to seven active slaves. This is because there two address types defined, such as: *Active Member Address (AMA)* on 3 bits that is valid as long as the device is an active member of a piconet. This means that we can have 8 active devices within a piconet out of which one is the master. *Parked Member Address* on 8 bits that is valid as long as the slave device is parked in a piconet [43].

The master coordinates the medium access and Bluetooth is using FHSS with a frequency hopping of 1600 hops/second while the hopping sequence is determined by the master device as a function of its Bluetooth address. All the devices in the piconet hop together and each piconet has an unique hopping sequence because one device cannot be the master in two piconets. Thus, the piconets have different masters and all the devices within the piconet need to

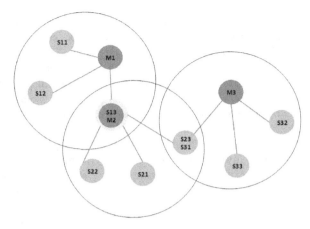

FIGURE 7.14
Bluetooth Architecture

be synchronized to the master's clock. Direct slave to slave communication is not possible and it has to go though the master. The medium access among different piconets is FH-CDMA. The piconet follows a centralized communication paradigm based on TDD where the master tells the slaves when to talk while the access method is TDMA as there are multiple devices sharing the piconet medium.

There are three classes of devices defined in Bluetooth, and a mixture of device classes can exist in one piconet. The device classes consist of:

- Class 1 – long range communications up to 100 m devices, maximum output power of 100 mW (20 dBm)

- Class 2 – ordinary range devices up to 10m, with a maximum output power of 2.4 mW (4 dBm)

- Class 3 – short range devices, up to 1m, maximum output power of 1 mW (0 dBm)

Bluetooth standard defines two physical links: (1) Synchronous Connection Oriented (SCO) similar to the circuit switched link from cellular systems, it is a voice link, full duplex, allocates fixed bandwidth between point-to-point connection of master and slave, 64 kbps, no re-transmissions supported; (2) Asynchronous Connection Less (ACL) similar to the packet switched link from cellular system, it is a data link, asynchronous, fast acknowledge, point-to-multipoint link between master and all slaves, up to 433.9 kbps symmetric or 723.2/57.6 kbps asymmetric.

Additionally, a Bluetooth device can be in several operational states, such as:

- standby – this is the default initial state where the device does nothing apart from waiting to join in a piconet

- inquire – search for other devices within range. In this state the Master sends an inquire packet. Slaves scan for inquiries and respond with their address and clock after a random delay (CSMA/CA).

- page – connect to a specific device. Master in page state invites devices to join the piconet.

- connected – the device participates actively in a piconet (master or slave).

- active state – master schedule transmissions

Apart from these operational states, there are three low power connected states defined for power saving modes.

- sniff – the device listens periodically the piconet at a reduced rate and maintains its AMA;

- hold – the device stops the ACL transmission but SCO is still possible and it also keeps its AMA, achieves less power consumption than sniff mode;

- park – the AMA is released and the device gets a PMA. The device wakes up at certain beacon intervals for re-synchronization; achieves less power consumption than hold mode.

As the technology evolved, different versions of the Bluetooth standard have emerged with different characteristics as listed below:

- **Bluetooth 1.1** also known as the IEEE Standard 802.15.1-2002 was the initial stable commercial standard

- **Bluetooth 1.2** also known as the IEEE Standard 802.15.1-2005 brought several enhancements, such as: eSCO (extended SCO) for higher, variable bitrates, asymmetric, re-transmission for SCO; and AFH (adaptive frequency hopping) to avoid interference;

- **Bluetooth 2.0 + EDR (2004, no more IEEE)** brought: EDR (enhanced date rate) of 3.0 Mbit/s for ACL and eSCO and lower power consumption due to shorter duty cycle

- **Bluetooth 2.1 + EDR (2007)** brought: better pairing support, e.g. using Near Field Communication (NFC) and also improved security;

- **Bluetooth 3.0 + HS (2009)** represents in fact Bluetooth 2.1 + EDR + IEEE 802.11a/g for data rate up to 54 Mbps;

- **Bluetooth 4.0 (2010) – Bluetooth Low Energy** – Low Energy, much faster connection setup;

- **Bluetooth 4.1** – enhancement of Bluetooth 4.0 + Core Specification Amendments (CSA) 1, 2, 3, 4;

- **Bluetooth 4.2 (Dec 2014)** – makes use of larger packets, improved security/privacy, and IPv6;

- **Bluetooth 5.0** brings data rates of up to 2 Mbps, 200 m LoS coverage, and support for IoT devices;

A new technology that has been introduced as part of the Bluetooth 4.0 specifications is the **BLE** which is not compatible with Bluetooth classic. It consumes less energy from 1% to 50% of the Bluetooth classic. The focus of BLE is more on IoT applications where small amounts of data are transferred at lower speeds up to 1 Mbps and does not support voice/video, file transfers. It makes use of the same 2.4 GHz band as Bluetooth classic, ZigBee, WiFi, etc. and it uses a star topology.

The BLE devices consist of *central devices* that are more capable in terms of CPU, power, memory or battery capacity and *peripheral devices* that are resource constrained devices especially in terms of battery. Thus, most of the processing is done by the central devices which allows the peripherals to sleep for longer or turn off their radio to consume less power. The BLE device discovery process consists of four types of advertisement packets:

- ADV_IND – used by peripheral devices to request connection to any central device.

- ADV_DIRECT_IND – connection request directed by the peripheral at a specific central device.

- ADV_NONCONN_IND – non connectable devices, advertising information to any listening device.

- ADV_SCAN_IND – Similar to ADV_NONCONN_IND, with optional additional information via scan responses.

Bluetooth 5.0 provides a mesh network via direct devices connections in a many-to-many topology in order to enable support for home automation and industrial applications.

With Bluetooth Mesh the aim is to extend the coverage area of Bluetooth networks in order to add support for more industrial applications. Thus, the devices can operate in the many-to-many topology and the device that is part of the mesh network is referred to as *node* while the devices that are not part of the network are called *unprovisioned* device. The unprovisioned device can get *provisioned* by joining the network and in this way it becomes a *node*. Within a node, there can be multiple parts that can be controlled independently and these are referred to as *elements*. The process for a device to become provisioned consists of six steps: (1) *Beaconing* – the unprovisioned device can trigger a mesh beacon advertisement by pressing a button for example; (2) *Invitation* – the provisioner will send out a *provisioning invite packet* to which the unprovisioned device responds with a *provisioning capabilities packet* that includes information about the number of elements the device supports, the set of security algorithms supported, the availability of its public key, etc. (3) *Public Key Exchange* – the provisioner and the unprovisioned device exchange public keys; (4) *Authentication* – the authentication of the unprovisioned device takes place based on the capabilities of both devices. For example, it can be through random number generation/input and confirmation generation/check; (5) *Provision Data Distribution* – each device derives a session key using their private key and the public key they have exchanged in step 3. The session key is then used to secure the connection for exchange of other additional provisioning data, which can include network key, device key, a security parameter, the unicast address allocated by the provisioner, etc. (6) *The unprovisioned device becomes a Node.*

The security for Bluetooth Mesh is mandatory and all the messages are encrypted and authenticated through three types of security keys addressing a different concern, such as: network security, application security and device security.

7.4 Introduction to Machine Learning for IoT

The aim of this section is to provide an overview of ML and to introduce the main concepts. More in-depth information on ML is provided by Lindholm et al. in [44] while more practical applications of ML can be found in [45].

The future of technology sees the fusion of the next generation networks, the IoT and AI to enable a so called intelligent connectivity [46] for new disruptive digital services. Intelligent connectivity is driven by IoT as an enabler of connecting any 'thing' to the Internet that will eventually generate data. All this Big Data that is collected from IoT has to be analysed and contextualized using AI techniques. However, in order to be useful for the user, it has to be presented in a meaningful way. Thus, the aim is to achieve a better interaction between people and the technological environment surrounding them through personalized delivery and improved users' QoE. This is done by improving the decision making and one important key enabling technology that can help to achieve this is the use of Machine Learning (ML).

The advancements in technology have changed the way we use and process the data. Due to IoT we generate a significant amount of data which we now have the ability to store and process. Furthermore, we can access this data over the Internet from remote locations. However, this data is relevant to us only when we can make something out of it and use it for example to make predictions [47]. Consequently, we need to analyze the data and turn it into something meaningful for the end-user. However, due to the heterogeneity and diversity of the data, this process is not straight forward. In this context, ML is seen as a key enabling technology that could provide the machines/systems the ability to optimize a Key Performance Indicator (KPI) using sample data or past experience. Thus, the aim of ML is to develop a computer program that can access and use training data or past experience to learn for themselves. ML is a subset of AI that makes use of the theory of statistics to build a mathematical model to enable a system to learn and adapt to changes in dynamic environments.

Consequently, ML should work in a similar way as the human brain. Even from birth it involves the process of constantly feeding information to the brain so that it can interpret it, understand the useful insides, detect patterns and identify key features to solve problems. For example, this is how a human has developed the capability to distinguish between two different objects. It continuously absorbed the information from the surroundings, collected the data and learnt from the past experiences. Consequently, will all this data the brain is capable enough to think and make decisions.

In general, we can identify three classes of ML, such as:

- **Supervised Learning** – in this type of ML, the machine learns under guidance provided by a supervisor. Thus, the supervisor is providing the labelled data and representing the correct output and the aim is to learn a mapping from the input to this output.

- **Unsupervised Learning** – in this type of ML, there is no supervisor and there is no correct output data provided either. Thus, the machine has to find regularities in the input data set given to be able to identify hidden patters in order to make predictions about the output.

- **Reinforcement Learning** – in this type of ML, the aim is to establish a pattern of behaviour through a sequence of actions. The machine follows the hit and trial concept, by interacting with an environment, producing actions and discovering errors/punishments or rewards and learns from this experience to be able to predict on the new data. Thus, in RL, the aim is the sequence of correct actions that enables the machine/system to reach the desired goal.

Consequently, *supervised learning* is a method through which the machine learns using labelled data provided by a supervisor, while in *unsupervised learning* the machine is trained on unlabelled data without any guidance from a supervisor, and *reinforcement learning* makes use of an agent that interacts with its environment by taking actions and transitioning from one state to the other state discovering errors or rewards, with the main goal to maximize the rewards.

These types of ML can be used to solve different types of problems. For example, the *supervised learning* is mainly used for two categories of problems, such as: *regression* and *classification*. There is an important difference between classification and regression, such that classification is about predicting a label or a class whereas regression is about predicting a continuous quantity. For example, a classic example is classifying our emails into two different groups. Thus, labelling the emails as spam and non-spam emails. For this kind of problem where we have to assign the input data into different classes we make use of classification algorithms. On the other hand regression is used to predict a continuous quantity representing a variable that has an infinite number of possibilities. Thus, regression is a predictive analysis used to predict continuous variables, when we do not have to label data into two different classes, but instead we need to predict a final outcome. For example, predicting the price of a stock over a period. Thus, supervised learning is widely used in the business sector for forecasting risks, risk analysis, predicting sales profit, etc.

On the other side, the *unsupervised learning* can be used to solve *association* and *clustering* problems. Association problems involve discovering patterns in the input data, by finding co-occurrences. A classic example of an association rule is by looking at the relationship between bread and jam. This concept is built on the idea that people do not go to the supermarkets to buy things randomly. Thus, looking into customer behavior we might see that customers who tend to buy bread also tend to buy jam. Thus, it is all about finding associations between items that frequently co-occur or items that are similar to each other. Apart from association problems unsupervised learning also deals with clustering and anomaly detection problems. Clustering is used for cases that involve for example targeted marketing wherein a list of customers and some information about them is given and a cluster of these customers based on their similarity has to be identified. Anomaly detection on the other hand is used for tracking unusual activities. A classic example is the credit card fraud where various unsupervised learning algorithms are used to detect suspicious activities in order to identify fraud. Thus, unsupervised learning is widely used for anomaly detection, credit card fraud detection, recommendation systems, etc.

The *reinforcement* learning is comparatively different from the rest and the key difference is that the input itself depends on the actions taken. Taking the classic example of a robot placed in an unknown environment. Thus, it starts in a situation where the robot does not know anything about the surrounding it is in. However, after it performs certain actions it finds out more about the environment but the environment it sees depends on whether it chooses to move right or whether it chooses to move forward or backward. In this case the robot is known as the agent and its surrounding is the environment so for each action it takes it can receive a reward or it might receive a punishment. However, its aim is to maximize the reward by taking a sequence of correct actions which would lead it to accomplish its goal. Thus, reinforcement learning is mainly used in self-driving cars and games development.

However, a machine will learn only after it was trained. Thus, in terms of training phase, in supervised leaning this is well defined and very explicit. The machine is given training data where both the input and output is labelled and the algorithm has to map the input to the output. Thus, the training data acts like the supervisor and once the algorithm is well trained it is tested using new data. When it comes to unsupervised learning the training phase takes longer because the machine is only given the input and it has understand the output on its own so there is no supervisor. In the case of reinforcement learning there is no predefined data and the whole reinforcement learning process itself is a training and testing phase. Since there is no predefined data given to the machine it has to learn everything on its own and it starts by exploring and collecting data from the environment.

As we can clearly see, the machine will learn faster with a supervisor. Thus, these types of ML algorithms are more commonly used in real-life tasks.

Some of the popular algorithms for supervised learning are *linear regression* which is mainly used for regression problems while algorithms like *support vector machines, decision trees, K-nearest neighbour*, etc. can be used for classification problems. For unsupervised learning we have algorithms like *K-means, C-means* for clustering analysis and algorithms like *Apriori* and *association rule mining* to deal with association problems. In the case of reinforcement learning a few algorithms include *Q-learning* and the *state action reward state action algorithm (SARSA)*.

Classification is the process used to separate the input data into two or more categories based on a known attribute/feature. Classification problems always require a supervisor. The data should be labelled with features so the machine could assign the classes based on them. Popular classification algorithms are: K-Nearest Neighbours, Decision Tree, Naive Bayes, Logistic Regression, Support Vector Machine.

Regression on the other hand is in fact classification where we forecast a number instead of a category. For example, the price of a car by its mileage, the traffic by time of the day, demand volume by growth of the company etc. Regression is perfect when something depends on time. Popular regression algorithms are: Linear and Polynomial regressions.

7.4.1 Linear Regression

Linear regression is used for solving regression problems and is used to estimate the relationship between variables with one variable dependent on one or more independent variables. For example, in case we want to predict the height of a person based on their weight, height would be the dependent variable and weight the independent one.

Figure 7.15 shows the relationship between the height on y-axis and weight on x- axis. This is linear regression, you use the line to predict the height given

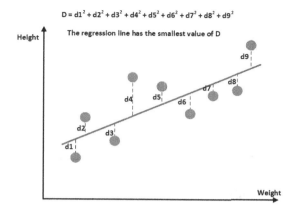

FIGURE 7.15
Linear Regression Example

the weight. Where D is the mean square error, representing the error in the prediction that indicates how much the predicted values represented by any point on the blue line, differ from the original values represented by the green dots. As it can be noticed, the distance of these data points from the line is the smallest relative to any other possible lines in other parts of the graph.

Thus, the concept behind linear regression is in fact fitting a line to data with least squares and R-squared.

7.4.2   K-Nearest Neighbours

In the K-Nearest Neighbors algorithm a new data point is assigned to a neighbouring group that is most similar to. This is done by taking a majority vote of its neighbours and assign the new data point to the most common of the categories among the voting neighbours. The number of nearest neighbours to consider is represented by K and can be defined by the user. This makes the number of neighbors K the core deciding factor. In general, K is defined as an odd number if there are only two classes. Furthermore, for K=1 the algorithm is known as the nearest neighbor algorithm.

Figure 7.16 illustrates an example of a simple KNN classifier. The new data to be classified would be classified as blue star if we take K=3 considering thus its 3 nearest neighbors, while if K = 7 then the new data would be classified as red triangle.

Consequently, in general there is no optimal number of neighbors that would suit all kind of data sets as each dataset has its own requirements. However, tn the case of a small value of K, the noise will have a higher influence on the results, while for a large value of K it will make the model computationally intensive.

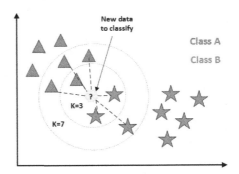

FIGURE 7.16
KNN Classifier Example

7.4.3 Decision Tree

The decision tree is a graph that makes use of a branching method to determine the class that the input needs to be assigned to. This method will illustrate every possible outcome of a decision. Figure 7.17 illustrates an example of a simple decision tree classifier. The decision tree consists of: a root node that represents the beginning of the decision tree; decision nodes that are represented by conditional control blocks that are split from the root nodes and further split based on other decision node; and the leaf nodes that are the final nodes representing the class label, the final decision taken after no further splitting is possible and after computing all the attributes.

7.4.4 Random Forest

A random forest classifier generates multiple decision trees this is because one single decision tree is not flexible enough when it comes to classifying new samples. Thus, decision trees only work great with the data used to create them. Consequently, the random forests are built from decision trees to combine their simplicity with the flexibility resulting in a significant improvement in accuracy when it comes to the classification of new data. There are several steps involved in creating a random forest, and the first step is to create a bootstrapped dataset from the original dataset but of the same size. Which means that we can randomly select samples from the original dataset that we put in our bootstrapped dataset allowing duplication so that we have the same size of the datasets. The next step would be to create a decision tree using the bootstrapped dataset, but only using a random subset of attributes at each step. This would result in a wide variety of trees that will help improve the accuracy. Once created, the random forest is used for new data that needs to be classified. This new data is run down each decision tree and the output of each decision tree is recorded. After the new data was run down all the trees

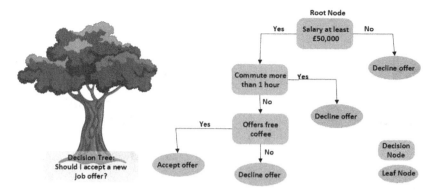

FIGURE 7.17
Decision Tree Classifier Example

in the random forest we check to see which output received the most votes. That will define the classification of the new data point.

Once we have created the random forest and we have used it, we want to validate it to see if it is working correctly. We do this by using the data entries that were not included in the bootstrapped dataset in the first step. This is called out-of-bag dataset. Thus, we run the out-of-bag dataset through the random forest and check to see if it correctly classified the samples.

7.4.5 Naive Bayes

Naive Bayes classifier is mostly used in cases where a prediction needs to be done on a very large data set. Thus, the complexity of the input dataset is high. It makes use of the conditional probability defined by Bayes' theorem as below:

$$P(A|B) = P(A)\frac{P(B|A)}{P(B)} \tag{7.1}$$

where $P(A|B)$ is the probability of event A occurring when B has already occurred, $P(A)$ is the probability of event A occurring, $P(B|A)$ is the probability of event B occurring when A has already occurred, and $P(B)$ is the probability of event B occurring.

This algorithm is most commonly used in filtering spam emails in your email account. However, it can be used to predict if on which days is possible to play cricket as illustrated in Figure 7.18. Based on the probability of a day being rainy, windy or sunny the model can tell us if a match is possible or not. If we consider all the weather conditions to be event B for us, and the probability of the match being possible event A., the model applies the probabilities of event A and C into the Bayes theorem and predicts if a game of cricket is possible on a particular day or not. In this case if the $P(A|B)$ is

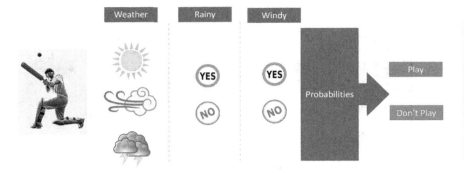

FIGURE 7.18
Naive Bayes Classifier Example

more than 0.5 we will be able to play a game of cricket if it is less than 0.5 we cannot.

7.4.6 Logistic Regression

We have seen previously that Linear Regression is used for solving regression problems while Logistic Regression is used for solving classification problems. Logistic regression is a classification algorithm, used when the value of the target variable is categorical in nature. Logistic regression is most commonly used when the data has binary output, meaning that it belongs to one class or another, thus it is either a 0 or a 1. Compared to linear regression, instead of fitting a line to the data, logistic regression fits an S shaped logistic function, known as sigmoid function and the curve obtained is called as sigmoid curve or S-curve as seen in Figure 7.19. The curve goes from 0 to 1 and it gives the probability that the output is either 0 or 1 based on the input data and using a threshold value. Thus, if we set the threshold value to 0.5 if the data point if below this threshold value it will be classified as a 0 and if it is above the threshold value it will be classified as a 1.

7.4.7 Cross-Validation

In general, when using ML algorithms for a given dataset, we need a part of the data to train the ML algorithms and another part to test the ML algorithms. If it was to use all the data to estimate the parameter and consequently to train the algorithm, we would not have any data left to test the model. We need to understand how the model will work on new data. In general, if we use the first 75% of the data for training and the last 25% of the data for testing is a better approach. However, it is not clear what is the best way to divide up the data, for example if using the first 25% of data for testing instead of the last or the 25% somewhere in the middle.

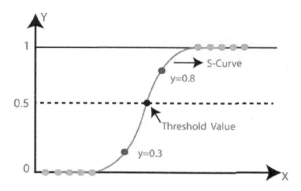

FIGURE 7.19
Logistic Regression Example

Consequently, the best approach would be to use them all through *cross-validation*. Cross-validation will use of the data, one at a time and summarizes the results at the end. The number of blocks the data is split into is arbitrary. However, in general, it is very common to divide the data into 10 blocks and this is called ten fold cross validation.

7.4.8 Bias and Variance

Two important concepts in ML are the bias and variance. Considering again a given dataset, first thing we do is to split the data into two sets, one for training the ML algorithms and one for testing them.

First we start with just the training set and we model a ML algorithm and we look at how accurately it can replicate the true relationship of the data. The inability for a ML method to capture the true relationship is called bias. For example, if a ML model cannot capture the true relationship it is said that it has a relatively large amount of bias. However, if another ML method might be very flexible and might fit exactly the training set along the true relationship, this is said to have very little bias.

However, apart from the training set there is also the testing set. The difference in fits between the training set and the testing set is called Variance. Thus, it might happen that the ML that had very little bias to do a terrible job at fitting the testing set which means that it has high variability. This means that with this kind of ML model it will be hard to predict how well it will perform with new datasets. It might do well sometimes and other times it might do terrible. In contrast, to the ML model with high bias it might have relatively low variance. Thus, this ML model might only give good predictions and not great predictions but they will be consistently good predictions.

7.4.9 Confusion Matrix

Similarly, given dataset, first thing we do is to split the data into two sets, one for training the ML algorithms and one for testing them. Then we train one or all the ML models we are interested in with the training data then test each ML model on the testing set and we need to summarize how well each method performed on the testing data. This can be done by using a *confusion matrix* as illustrated in Figure 7.20

For example, if we have two categories to choose from like Positive or Negative then the top-left corner contains true positives these are the data entries that were correctly identified as Positive by the algorithm, the true negatives are in the bottom right hand corner these are data entries that were correctly identified as Negative by the algorithm, the bottom left-hand corner contains false negatives which the ones where the actual data entry was Negative but the algorithm classified as Positive, and the top right-hand corner contains false positives which are represented by the data entries that where actually Positive but the algorithm classified them as Negative.

		Predicted Values	
		Positive	Negative
Actual Values	Positive	True Positive (TP), hit	False Negative (FN), Type II error, miss, underestimation
	Negative	False Positive (FP), Type I error, false alarm, overestimation	True Negative (TN), correct rejection

FIGURE 7.20
Confusion Matrix Example

This example is just a 2x2 matrix because we had only two categories to choose from. However, if there are more categories to choose from, then our confusion matrix will have more rows and more columns.

7.5 Digital Twins for Industrial IoT

The rapid advancements in manufacturing technologies are transforming the current industrial landscape through Industry 4.0, which refers not only to the integration of information technology with industrial production but also to the use of innovative technologies and novel data management approaches. The target is to enable the manufacturers and the entire supply chain to save time, boost productivity, reduce waste and costs, and respond flexibly and efficiently to consumers' requirements. Industry 4.0 moves the digitization of manufacturing components and processes a step further by creating smart factories. This is done by taking on board the ICT for evolution in supply chain and production line [48]. Within this context, one of the key enabling technologies for Industry 4.0 is the Digital Twin (DT). Moreover, the advancements in technologies along with the use of IoT, uRLLC, ML, Transfer Learning (TL), AI, data augmentation, cloud services, visualization, and Cloud, Edge & Fog computing enables several new venues and application areas for DT. Looking at the current leading vendors within the smart manufacturing Industrial Internet of Things (IIoT) environment, the creation of so called DT has become more and more of a common practice. However, looking into more details there is no common understanding of the definition of the DT between the leading vendors [49], and its usage is slightly different but showcased under the same umbrella of DT. It was noticed that terms like digital model, digital shadow, digital thread as well as digital twin are used synonymously [50]. However, there is a significant difference between these terms and this is mainly related to the level of interaction between the physical object and the digital object. The digital twin would represent the two-way interaction between the twin entities. However, a digital model represents just the digital representation of the physical object without any form of interaction between the physical object and the digital counterpart. Whereas a digital shadow would only enable a one-way communication link between the physical object and the digital object, where changes in the physical object would be reflected in the digital object but not the other way around.

A working definition of DT is given as follows: a DT represents the interconnection and convergence between a physical system/physical object and its digital representation created as an entity of its own [50]. However, to form a DT, both entities, the physical object and the digital object need to be fully integrated and exchange information in both directions. Thus, it is expected that the digital object should be able to act as a controlling instance of the physical object and vice versa. Within the DT context, IoT is used to automatically collect the manufacturing data in real-time while the DT along with big data analytics could use the data to predict, estimate and analyse the dynamic changes within the physical object. An optimized solution is then fed back to the physical object that would adapt accordingly [49]. To this end, this makes the DT technology the main focus of the global manufacturing transformation as it has the potential to optimize the whole manufacturing process [51]. Within this context, a set of key challenges could be identified when creating a DT, as follows:

- physical entities represented by digital twins will require multiple models addressing different perspectives of the physical entity. Model integration is difficult because of a lack of a shared semantics between types of models.

- digital twin representations offered by manufacturers create platform dependence. Future smart applications of domains will require deployment of multiple technologies from different manufacturers. A common reference model for DT model representation is necessary and is not yet available.

- data analytics through integrated machine learning will be corner stone for use cases of digital twins. Seamless integration of ML techniques that are built on a common reference model is not yet available.

- DTs are on an evolutionary journey. Intermediate states will necessarily require human intervention. Modelling of human interactions as part of a digital twin and its usage will require novel methodological techniques.

- methodologies and patterns of digital twin deployment need to be developed concurrently with technology advances in DT.

- DTs, their development and deployment become a sociotechnical innovation with attendant risks associated with how DTs may diffuse or embed certain moral values. For example, how can we assure of the transparency of decision making enacted through a DT and its analytical tools? The problem is compounded by the use of ML algorithms [52].

Consequently, most of the DT solutions provided by the current leading vendors are in fact digital models or digital shadows and not digital twins. This is because, creating a fully integrated DT is a challenging task due to the high and complex configuration as outlined above. The DT needs to be fully integrated within the Product Line Engineering (PLE) and interact with the environment and its physical processes. This integration becomes even more complex because of the heterogeneity of components and the tight interaction between software, networks/platform and physical components. Thus, most of the existing DT solutions provide only one way of communication between the physical object and the digital object, making them digital shadows. For these systems, to enable the two way communication and give birth to the DT the human action is required, and we refer to this as human in the loop.

7.5.1 DT for Manufacturing

Creating DTs within the manufacturing environment is becoming more and more of a common practice. However, because there is no common definition of the DT concept adopted globally its usage has a slightly different meaning between manufacturers, even though it is presented under the same umbrella of DT. Currently, most of the implementations of the DT concept relay in fact on the definition that a DT represents a high fidelity digital replica of the physical system/object without or with only one level of interaction from the physical object to the digital object. Thus, the changes in the physical object would be reflected into the digital object but not the other way around. This is mainly due to the significant challenges faced because of the heterogeneity of components, high and complex configuration and interactions.

However, the continuous industrial adoption of DT technology represents a great potential to reshape the future across diverse industries leading towards the Industry 5.0 revolution that sees the elevation of the level of virtual interactions with the physical environment. We have seen that the DT concept goes beyond the traditional computer-based simulations and analysis and it represents a two-way communication bridge between the physical world and the digital world. The physical object exists in symbiotic relationship with its digital counterpart, being connected through real-time data communications and information transfer. The existence of DTs is enabled by the advances in IIoT, AI, AR, VR, Big Data Analytics, etc. An example of fully integrated DT solution that enables real-time two-way communication between the physical object and the digital object for product service enhancement and optimized operations, is illustrated in Figure 7.21 [53].

Figure 7.21 illustrates the DT framework of a Festo Cyber-Physical (CP) Factory installed at Middlesex University [54] which is a dedicated system that performs the production assembly line operations. The CP Factory consists of two production units referred to as islands that are interconnected via an Automated Guided Vehicle (AVG). There are fours stations on each island that perform dedicated assembly tasks and an abridged station that handles the passing of the product to the AVG for transportation to the other island. An order can be placed through the Manufacturing Execution System (MES)

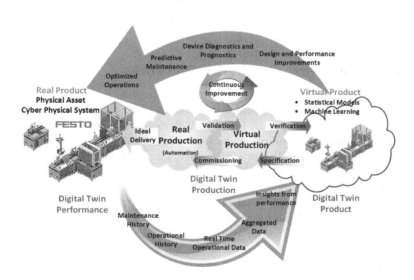

FIGURE 7.21
DT Framework

and a carrier tray is assigned to that particular order for the entire duration of the process. A conveyor belt will move the carrier tray from station to station with each station being able to identify the carrier, order number and progress using RFID to ensure that every order is processed only once [55].

The first step in developing the DT of the CP Factory is to create its 3D model which was implemented in Unity, a game engine with a C#-based API. Once the 3D model of the system had been created the next step is to establish the two-way communication between the physical and digital twin. Consequently, this is done by connecting the sensors at each station from the real CP Factory to its digital model. Thus, the data coming from capacitive sensors installed on the conveyor, along with the optical encoder measuring motor speed, is continuously sent to the digital twin. The successful implementation of this two-way connection betwenen the physical and digital twin, enables harnessing the predictive maintenance capabilities of DTs to increase production efficiency. Additionally, the DT can leverage the digital replica's operation history and simulation capabilities to achieve accurate and adaptive predictive maintenance operations [54].

The aim of smart manufacturing is to make operations flexible, adaptable and optimizable within an ecosystem of coexistence between people, processes and machines. This is achieved, by integrating various Industry 4.0 technologies (AI, Big Data Analytics, B5G, Edge/Cloud Computing, IIoT, etc.) and extending the capabilities of both manufacturing devices and people to optimize efficiency and productivity within smart factories. Within this context, the future vision is to go beyond the simple notion of individual DT and focus on developing an intelligent interconnected ecosystem of DTs to help enable smart manufacturing across factory and supply chain as illustrated in Figure 7.22.

7.5.2   DT for Next Generation Networks

The DT technology has been used in many industries, including manufacturing, energy, industrial assets and structures. A dual fault diagnosis method based on DT was developed in [56] which demonstrates high diagnosis accuracy in predicting the trend of production throughput with respect to changes in working conditions and data efficiencies. The work in [57] proposed a DT model for fault diagnosis of rotation machinery, unifying physical knowledge, experimental data, and model updating technique to achieve a clear improvement compared to the traditional fault diagnosis method with error rates under 5% in locating fault and assessing its extent. The work in [58] presented a DT-based real-time monitoring system for mechanical structures to improve the safety of the work environment using IoT and augmented reality. Liao et al. [59] presented a cloud-based open-source framework named SnowFort using a wireless sensor framework for the monitoring of infrastructure systems. The

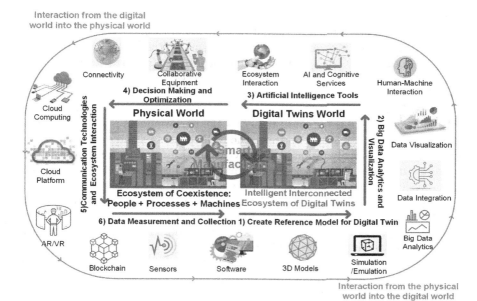

FIGURE 7.22
DTs Ecosystem Vision

Predix platform of General Electric[2] has capability of ingesting large volumes of sensory data, running analytic models, and performing business rules at the same time, allowing detection of abnormal phenomenons and improve plant reliability. Siemens empower their friendly 3D CAE software Simcenter [60] with DT approach by combining physics-based simulation, closed-loop of data from operation to design, and their IoT system MindSphere, to perform real-time simulation and achieve more predictive results throughout the product life cycle. A number of case studies have been realized, such as forecasting the deformation of a blade of Gas Turbines in extreme environmental conditions, improving user's thermal comfort in electric vehicles by balancing the HVAC system and battery performance, optimizing the efficiency of excavator through a co-simulation of excavator and soil condition.

However, most of these applications of DTs rely on the underlying infrastructure provided by the next generation networks in order to meet the requirements of their strict QoS.

Looking at the current next generation networks (NGN) environment we can notice there are several challenges that the network operators are facing in terms of how to speed up the deployment of new (but complex) NGN technologies; how to provide flexible test-bed facility with high availability; and willingness to invest in the expensive NGN development with uncertain

---

[2]https://www.ge.com/digital/

returns. This has led to the concept of having the DT of the network itself [61] that can accurately replicate the entire next generation networks ecosystem and help tackle all the above obstacles to satisfy the NGN advancements needs. Using NGN DT has gained significant interest from the leading telcos, e.g., Ericsson [62], Huawei [63] where sensor/network data, traffic data, data mining, data visualization, and data interpretation are integrated into one system to facilitate the live replica of a process or whole NGN network. The NGN DT certainly has potential to access the performance, predict impact of the environment change and optimize the NGN processes and decision making accordingly. Thus a cloud-based DT NGN could aim to perform continuous designing, monitoring and proactive maintenance through the closed-loop data from physical entities to virtual counterparts and vice-versa.

The benefits of having a DT NGN are significant for many industries and this is because it enables them to build what-if scenarios for new products/services/processes or new network entities and test them under realistic conditions before moving into real-world implementation. The NGN DT could also improve ongoing operations by continuously monitoring the real physical systems and use Big Data analytics and ML to predict any issues before they would happen in real world.

This shows that the potential of the NGN DT technology and beyond is limitless. The implementation and automation of *IIoT* could benefit from NGN DT to reduce maintenance issues and optimize the production. Similarly, NGN DT could revolutionize *healthcare* operations as well as it could provide solutions for intelligent *assisted living homes*. The future of NGN DT could also envision building AI-powered instructors, teachers, doctors, nurses, etc.

Even though within the current environment the mobile operators are still working at deploying the 5G networks, in the research area works are carried out towards defining the next 6G networks [64]. The idea behind 6G is to enhance even further all the applications and vertical use cases of the 5G network by bringing the intelligence at the edge of the network. Consequently, 6G will bring an intelligent interconnected system of DTs that enables the creation of a real-time digital world. Thus, 6G will be represented by connected and augmented intelligence that will change the way the data is created, processed and consumed [61].

7.6 Use-Case Scenario: Technology for Public Health Emergencies

The advancements in technology together with the affordability of mobile devices open up new opportunities when responding to public health emergencies. The integration of technology in emergency response could represent a dramatic shift when dealing with public health interventions by enabling a faster coordinated response. The significant growth in the number of users with mobile phones as well as the adoption of key enabling technologies like cloud computing led to the creation of an entire tracking ecosystem that could enable the use of various surveillance methods. However, in this context a highly relevant issue is the data privacy, as the current governance and regulation frameworks are lagging behind all these technological advancements.

This was visible in the recent COVID-19 global pandemic situation, where concerns around privacy and civil liberties led to various countries to respond differently in an effort to control the spread of COVID-19 and preserve human life.

In this context, there is a proliferation of pandemic-tracking apps that try to get accurate access to the mobile phone's GPS location and other personal data. While some countries adopt a forced mass surveillance method that limits individual freedoms, some other countries rely on the voluntary adoption of the pandemic-tracking apps that try to get the population to consent to sharing their location and other personal data. The attempt to fight the pandemic could actually test people's attitude towards privacy and government surveillance [65–70].

Apart from large-scale testing, South Korea, Israel and China were the first to adopt mass surveillance contact-tracing systems in an attempt to quickly identify the exposed population and suppress the spread of the virus. Learning from the recent experience with the Middle East Respiratory Syndrome (MERS) Coronavirus outbreak in 2015, South Korea's population accepted that a privacy trade-off is required. Thus, GPS, CCTV footage, credit card transaction, and travel information data was used by the epidemiological intelligence officers to monitor the population and make sure those infected or quarantined obey the rules otherwise they will be punished with a location-tracking bracelet or even incarceration [71].

In Israel, the government sought the assistance of the Israeli Security Agency (ISA) to work together with the Health Ministry under the Coronavirus Location Tracking Temporary Provisions to contain the spread of the virus. The type of shared data includes: name, identification number, phone number, date of birth, cell phone location data, traffic history (e.g., timestamp and phone numbers of all parties involved in a phone call). However, there are some constraints imposed on the duration of data collection and storage. Consequently, ISA can store the data over a period of 14 days, while the Health Ministry can retain the data up to 60 days.

China made use of its extensive surveillance infrastructure to contain the spread of the virus without much consideration of the individual privacy rights. Health QR codes were embedded in popular mobile phone apps that generate a rating indicating the health status of an individual and their likelihood of having contacted the virus based on their travel and medical data. In case a person tests positive, the authorities will release public data including the person's address and movement history. Surveillance footage along with facial recognition and movement mobile phone data are used to monitor any quarantine violations which are then associated with severe penalties.

These centralized mass surveillance methods enforced in South Korea, Israel and China along with mass testing have been effective in containing the spread of the virus but at the cost of the population's privacy rights. On the other side, in Europe, the European Union General Data Protection Regulation (GDPR) has more protective privacy laws on the collection, use, and storage of personal digital data, that forced the European countries to look into developing contact-tracing applications that work based on the voluntary adoption of the population. The contact tracing applications work on proximity data indicating the likelihood of someone being infected based on the epidemiological distance and duration of contact with an infected person. Although, for this approach to be successful, it requires high adoption and engagement from the population which is unlikely to achieve due to the general privacy concerns. However, even in countries where data privacy is not formally at risk, citizens may have a different perception of what they think as legally permissible or safe from a fundamental rights perspective. From a socio-legal perspective, this phenomenon can be regarded as a discrepancy between formal legality and legal reality.

In an attempt to overcome the security and privacy concerns, various COVID-19 tracing applications adopted different architectures for data collection. Consequently, these architectures could be classified into three categories [72]: (1) centralized, (2) decentralized and (3) hybrid which combines the features of the centralized and the decentralized approached. The classification is done based on how the server is used and the type and location of the data.

Centralized Approach

Within the centralized architecture most of the information is stored and processed on one cloud server. In this approach, the cloud server plays a key role by storing pseudonymous users' personal information, performing risk analysis and sending out notifications to close contacts in case of infection. Consequently, this raises security and privacy concerns around the use and the life cycle of the data collected especially if the cloud server becomes an untrustable entity. The available information on the centralized server can be further used for data analysis that could help the government deciding on lockdown restrictions in hot-spot areas.

Decentralized Approach

Within the decentralized architecture, the core functionalities are moved to the user devices and the involvement of the centralized server within the contact tracing process is drastically reduced. This approach tries to enhance the users' privacy by performing the tracing process locally on the user's device. The contact tracing applications based on the decentralized approach do not require the users to pre-register before use. Thus, there is no personally identifiable Information stored on the server. A device running the app will generate privacy-preserving pseudonyms that are exchanged periodically between the devices that come in close contact. The central cloud server in this scenario acts as a rendezvous point for lookup purposes where the infected user can volunteer to upload their relevant time information which reflect only their trajectory and does not include any information about the encounters. The other app users can pull this information regularly from the central server to be used locally on their devices to perform a risk analysis to check if they have been exposed to the virus and for how long. Even though this approach alleviates some of the privacy risks, there is no information on the central server for data analysis purposes that could help the government deciding on lockdown restrictions in exposed hot-spot areas.

Hybrid Approach

Ahmed et al. [72] define the Hybrid approach as a combination of the centralized and decentralized approaches. Within the hybrid architecture the centralized cloud server does not send the encounter information from the infected users to others and the risk analysis and the notifications are handled locally at the server. This approach avoids the user de-anonymisation attacks that could happen within the decentralized approach. In this case, an infected user voluntarily uploads the required data to the centralized server. Other users could check their risk exposure by inquiring the server which computes the risk analysis and notifies the users in case they need to contact the health authority. The advantage of this architecture is that statistical information is available at the server that could be used to identify exposure hot-spots and deciding on the required measures depending on the pandemic circumstances. However, most of the related surveys in the literature [73–76] classify the contact tracing solutions into two categories only, such as centralized and decentralized as illustrated in Figure 7.23

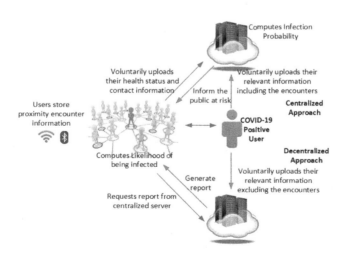

FIGURE 7.23
Centralized vs. Decentralized Approach of Contact Tracing Apps

In terms of the technology being used, both the centralized and decentralized approaches mainly rely on Bluetooth, Global Positioning System (GPS), Quick Response (QR) codes and cellular location tracking [73]. A summary of the contact tracing apps adopted by different countries around the world, including their names and their corresponding technologies is illustrated in Figure 7.24. Any of the technologies used has its own advantages and disadvantages in terms of accuracy, privacy concerns and data protection. However, regardless of the technology being used, the main technical requirements of any contact tracing app is that it must operate at close range to be able to determine with high accuracy if a person has been within the 2m proximity of an infected individual. While GPS could provide a location accuracy between 10 to 20 meters only, cellular location data is even less precise and adds significant privacy concerns that could violet citizens' data protection rights. Consequently, most of the contact tracing apps rely on Bluetooth which operates at close range and has a reasonable accuracy within the 2 m proximity. Moreover, the individual still holds the control and could decide to opt in or out by switching the Bluetooth on or off. As seen in Figure 7.24, most countries rely on Bluetooth as the best choice regardless of the approach implemented.

FIGURE 7.24
Centralized vs. Decentralized Contact Tracing Apps' Approaches Adopted Around the World and the Technologies Involved

Countries like France, Australia, Singapore, New Zealand, Norway, India, Mexico, Qatar, Kuwait, Bahrain, Hungary, Bulgaria, Tunisia have opted for centralized approaches. The centralized apps mainly follow the PEPP-PT (Pan-European Privacy-Preserving Proximity Tracing) protocol [77]. Most of the centralized approaches combined Bluetooth with location information to improve the accuracy. However, in Norway the Data Protection Authority has suspended the app on the grounds that poses a significant threat to user privacy by continuously uploading individuals' location information. The UK initially adopted a centralized approach, but due to privacy concerns and mobile devices battery drainage, it switched to a decentralized solution. Belgium, Switzerland, Finland, Estonia adopted the decentralized approach that follows the DP-3T (Decentralized Privacy-Preserving Proximity Tracing) [78] which is seen as a partially decentralized solution as it uses an anonymous centralized database for the infected individual. However, the identification of a specific individual is not possible through the type of data collected and exchanged. Recently, most of the countries adopted the decentralized approach that relies on the cross-platform API developed by Google and Apple [79]. However, despite the improvements around privacy and security, there are still concerns that Google/Apple could end up controlling the EU's Covid-19 app ecosystem.

Despite all the efforts across the world, it is obvious that finding the balance between the use of an effective technology-based contact tracing application and the data protection and privacy of individuals remains a challenge. The main concerns around the two approaches (centralized vs. decentralized) are related to the type and location of the data collected (e.g., if this is held by the government or it remains with the users). The countries that prefer to retain control opted for the centralized approach as in this way, the government can impose the required steps to control the spread of the virus instead of relying on the users to act on the information provided by the app. However, a comparison between the efficiency of the two approaches centralized vs. decentralized and their contribution towards slowing down the spread of Covid-19 is difficult to make due to the high number of factors involved (number of participants, population density, running duration [e.g., how long was the app running in that particular country], etc.). Despite the fact that an actual comparison on the impact of either approach is not possible, integrating a contact tracing app is not worse than the scenario of not having an app.

7.7 Practical Use-Case Scenario: ML for Predictive Maintenance and IoT Using Python, Tensorflow, Jupyter

This practical use-case scenario is based on Miniconda[3] release of Python which already includes most of the required packages. At the time of the writing this use-case scenario was run using Python 3.6.10, TensorFlow Version 2.3, Keras Version 2.4.0 and Jupyter Notebook version 6.1.1.

The aim of this use-case scenario is to see how we can use ML for Predictive Maintenance [45]. The code for *Predictive Maintenance using Long Short-*

---

[3]https://docs.conda.io/en/latest/miniconda.html#windows-installers

Term Memory (LSTM) has been implemented by Umberto Griffo [80] and is available on GitHub.[4]

7.7.1 Use-Case Scenario Settings

In order to demonstrate the predictive maintenance, we are going to use simulated aircraft sensor values that are provided in Azure ML[5]. Based on this data, the model will need to predict when an engine of the aircraft is going to fail so that maintenance can be planned prior to this event. The aim is to understand if we could predict when an in-service engine will fail. In order to achieve this, two different approaches can be taken using machine learning models, such as:

- **binary classification**: could help us predict if the in-service engine will fail within certain time frame in terms of understanding if the engine will fail within a certain number of cycles.

- **regression models**: could help us predict the *Remaining Useful Life (RUL)* or *Time to Failure (TTF)* in terms of the number of cycles an in-service engine still has before it fails.

An assumption considered for this use case scenario is that the engine has a progressing degradation pattern reflected in the sensor measurement data. Consequently, we would try to predict future failures by monitoring the engine's sensor values over time and using ML to learn the relationship between the sensor values and the variation in sensor values as compared to the historical failures.

7.7.2 Data Sets Summary

This use-case scenario takes three datasets as inputs, freely available on Azure ML, as: PM_train.txt[6], PM_test.txt[7], and PM_truth.txt[8]:

- Training data: consists of the readings from 21 sensors for each cycle as the time unit from 100 unique engines ID. This data set represents the engine's run-to-failure data. However, it is assumed that the initial wear and manufacturing variation is different for each engine to start with. Moreover, at the start of each time series the engine is assumed to be operating normally but starts to progressively degrade during the operating cycles until a predefined threshold is reached and the engine is considered unsafe for operation. Thus, the failure point for each engine is indicated by the last cycle in each time series.

---

[4]https://github.com/umbertogriffo/Predictive- Maintenance- using- LSTM
[5]https://gallery.azure.ai/Collection/Predictive-Maintenance-Template-3
[6]http://azuremlsamples.azureml.net/templatedata/PM_train.txt
[7]http://azuremlsamples.azureml.net/templatedata/PM_test.txt
[8]http://azuremlsamples.azureml.net/templatedata/PM_truth.txt

- Testing data: consists of the same structure as the training data. However, the without the record of failure events.

- Ground truth data: consists the number of the true remaining working cycle for the engines in the testing data set.

7.7.3 Source Code and Results

This subsection explains the main building source code blocks used to solve this problem using the two approaches: binary classification and regression. For both approaches the Long Short-Term Memory (LSTM) is used [45].

Binary Classification – which predicts weather the engine will fail in a certain time frame (e.g., cycles).

1. First step would be to import all the modules required to implement the predictive maintenance and a seed is set for reproductibility.

```
1  import keras
2  import pandas as pd
3  import numpy as np
4  import matplotlib.pyplot as plt
5  import os
6
7  # Setting seed for reproducibility
8  np.random.seed(1234)
9  PYTHONHASHSEED = 0
10
11 from sklearn import preprocessing
12 from sklearn.metrics import confusion_matrix, recall_score,
   ↪  precision_score
13 from keras.models import Sequential,load_model
14 from keras.layers import Dense, Dropout, LSTM
```

2. Data Ingestion – the three data sets are read and the columns are assigned names:

```
1  # define path to save model
2  model_path = 'binary_model.h5'
3
4  # read training data - It is the aircraft engine
   ↪  run-to-failure data.
5  train_df = pd.read_csv('PM_train.txt', sep=" ",
   ↪  header=None)
6  train_df.drop(train_df.columns[[26, 27]], axis=1,
   ↪  inplace=True)
```

```
7   train_df.columns = ['id', 'cycle', 'setting1', 'setting2',
    ↪  'setting3', 's1', 's2', 's3',
8                            's4', 's5', 's6', 's7', 's8', 's9',
                         ↪  's10', 's11', 's12', 's13', 's14',
9                            's15', 's16', 's17', 's18', 's19',
                         ↪  's20', 's21']
10
11  train_df = train_df.sort_values(['id','cycle'])
12
13  # read test data - It is the aircraft engine operating data
    ↪  without failure events recorded.
14  test_df = pd.read_csv('PM_test.txt', sep=" ", header=None)
15  test_df.drop(test_df.columns[[26, 27]], axis=1,
    ↪  inplace=True)
16  test_df.columns = ['id', 'cycle', 'setting1', 'setting2',
    ↪  'setting3', 's1', 's2', 's3',
17                           's4', 's5', 's6', 's7', 's8', 's9',
                         ↪  's10', 's11', 's12', 's13', 's14',
18                           's15', 's16', 's17', 's18', 's19',
                         ↪  's20', 's21']
19
20  # read ground truth data - It contains the information of
    ↪  true remaining cycles for each engine in the testing
    ↪  data.
21  truth_df = pd.read_csv('PM_truth.txt', sep=" ",
    ↪  header=None)
22  truth_df.drop(truth_df.columns[[1]], axis=1, inplace=True)
```

3. Data Preprocessing (training data set) – a new label named Remaining Useful Life (RUL) is created and another label 'label1' is a binary variable that indicates if the specific engine is going to fail within w1 cycles. The non-sensor data is also normalized.

```
1   #######
2   # TRAIN
3   #######
4   # Data Labeling - generate column RUL(Remaining Usefull
    ↪  Life or Time to Failure)
5   rul = pd.DataFrame(train_df.groupby('id')['cycle'].max()).
    ↪  reset_index()
6   rul.columns = ['id', 'max']
7   train_df = train_df.merge(rul, on=['id'], how='left')
8   train_df['RUL'] = train_df['max'] - train_df['cycle']
9   train_df.drop('max', axis=1, inplace=True)
10
```

```
11  # generate label columns for training data
12  # we will only make use of "label1" for binary
    ↪   classification,
13  # while trying to answer the question: is a specific engine
    ↪   going to fail within w1 cycles?
14  w1 = 30
15  w0 = 15
16  train_df['label1'] = np.where(train_df['RUL'] <= w1, 1, 0 )
17  train_df['label2'] = train_df['label1']
18  train_df.loc[train_df['RUL'] <= w0, 'label2'] = 2
19
20  # MinMax normalization (from 0 to 1)
21  train_df['cycle_norm'] = train_df['cycle']
22  cols_normalize =
    ↪   train_df.columns.difference(['id','cycle','RUL',
    ↪   'label1','label2'])
23  min_max_scaler = preprocessing.MinMaxScaler()
24  norm_train_df =
    ↪   pd.DataFrame(min_max_scaler.fit_transform(train_df
    ↪   [cols_normalize]),
25                                  columns=cols_normalize,
26                                  index=train_df.index)
27  join_df =
    ↪   train_df[train_df.columns.difference(cols_normalize)].
    ↪   join(norm_train_df)
28  train_df = join_df.reindex(columns = train_df.columns)
29
30  train_df.head()
```

4. Data Preprocessing (test data set) – similar to the data preprocessing on
   the training data set with the difference that the RUL value is obtained
   from the ground truth data.

```
1  ######
2  # TEST
3  ######
4  # MinMax normalization (from 0 to 1)
5  test_df['cycle_norm'] = test_df['cycle']
6  norm_test_df =
   ↪   pd.DataFrame(min_max_scaler.transform(test_df
   ↪   [cols_normalize]),
7                                  columns=cols_normalize,
8                                  index=test_df.index)
```

```
9   test_join_df =
    ↪   test_df[test_df.columns.difference(cols_normalize)].
    ↪   join(norm_test_df)
10  test_df = test_join_df.reindex(columns = test_df.columns)
11  test_df = test_df.reset_index(drop=True)
12
13
14  # We use the ground truth dataset to generate labels for
    ↪   the test data.
15  # generate column max for test data
16  rul = pd.DataFrame(test_df.groupby('id')['cycle'].max()).
    ↪   reset_index()
17  rul.columns = ['id', 'max']
18  truth_df.columns = ['more']
19  truth_df['id'] = truth_df.index + 1
20  truth_df['max'] = rul['max'] + truth_df['more']
21  truth_df.drop('more', axis=1, inplace=True)
22
23  # generate RUL for test data
24  test_df = test_df.merge(truth_df, on=['id'], how='left')
25  test_df['RUL'] = test_df['max'] - test_df['cycle']
26  test_df.drop('max', axis=1, inplace=True)
27
28  # generate label columns w0 and w1 for test data
29  test_df['label1'] = np.where(test_df['RUL'] <= w1, 1, 0 )
30  test_df['label2'] = test_df['label1']
31  test_df.loc[test_df['RUL'] <= w0, 'label2'] = 2
32
33  test_df.head()
```

5. LSTM for time-series modelling – defines a window size of 50 cycles to be used in a function that will generate the sequence representing the input into the LSTM as per the respective window size. Another function will generate the corresponding label for the data.

```
1   # pick a large window size of 50 cycles
2   sequence_length = 50
3
4   # function to reshape features into (samples, time steps,
    ↪   features)
5   def gen_sequence(id_df, seq_length, seq_cols):
6       """ Only sequences that meet the window-length are
        ↪   considered, no padding is used. This means for
        ↪   testing
```

```
 7      we need to drop those which are below the
    ↪   window-length. An alternative would be to pad sequences
    ↪   so that
 8      we can use shorter ones """
 9      # for one id I put all the rows in a single matrix
10      data_matrix = id_df[seq_cols].values
11      num_elements = data_matrix.shape[0]
12      # Iterate over two lists in parallel.
13      # For example id1 have 192 rows and sequence_length is
    ↪    equal to 50
14      # so zip iterate over two following list of numbers
    ↪    (0,112),(50,192)
15      # 0 50 -> from row 0 to row 50
16      # 1 51 -> from row 1 to row 51
17      # 2 52 -> from row 2 to row 52
18      # ...
19      # 111 191 -> from row 111 to 191
20      for start, stop in zip(range(0,
    ↪   num_elements-seq_length), range(seq_length,
    ↪   num_elements)):
21          yield data_matrix[start:stop, :]
22
23  # pick the feature columns
24  sensor_cols = ['s' + str(i) for i in range(1,22)]
25  sequence_cols = ['setting1', 'setting2', 'setting3',
    ↪   'cycle_norm']
26  sequence_cols.extend(sensor_cols)
27
28  # generator for the sequences
29  seq_gen = (list(gen_sequence(train_df[train_df['id']==id],
    ↪   sequence_length, sequence_cols))
30              for id in train_df['id'].unique())
31
32  # generate sequences and convert to numpy array
33  seq_array =
    ↪   np.concatenate(list(seq_gen)).astype(np.float32)
34  print(seq_array.shape)
35
36  # function to generate labels
37  def gen_labels(id_df, seq_length, label):
38      # For one id I put all the labels in a single matrix.
39      # For example:
40      # [[1]
41      # [4]
42      # [1]
```

```
43    # [5]
44    # [9]
45    # ...
46    # [200]]
47    data_matrix = id_df[label].values
48    num_elements = data_matrix.shape[0]
49    # I have to remove the first seq_length labels
50    # because for one id the first sequence of seq_length
      ↪ size have as target
51    # the last label (the previus ones are discarded).
52    # All the next id's sequences will have associated step
      ↪ by step one label as target.
53    return data_matrix[seq_length:num_elements, :]
54
55  # generate labels
56  label_gen = [gen_labels(train_df[train_df['id']==id],
      ↪ sequence_length, ['label1'])
57              for id in train_df['id'].unique()]
58  label_array = np.concatenate(label_gen).astype(np.float32)
59  print(label_array.shape)
```

6. Building the LSTM model and train it. The LSTM model consists of two stacked LSTM layers with 100 and 50 units, respectively and a fully connected layer. The LSTM model makes use of the Adam optimizer to update the model parameters.

```
1   # Next, we build a deep network.
2   # The first layer is an LSTM layer with 100 units followed
      ↪ by another LSTM layer with 50 units.
3   # Dropout is also applied after each LSTM layer to control
      ↪ overfitting.
4   # Final layer is a Dense output layer with single unit and
      ↪ sigmoid activation since this is a binary
      ↪ classification problem.
5   # build the network
6   nb_features = seq_array.shape[2]
7   nb_out = label_array.shape[1]
8
9   model = Sequential()
10
11  model.add(LSTM(
12          input_shape=(sequence_length, nb_features),
13          units=100,
14          return_sequences=True))
15  model.add(Dropout(0.2))
```

```
16
17  model.add(LSTM(
18           units=50,
19           return_sequences=False))
20  model.add(Dropout(0.2))
21
22  model.add(Dense(units=nb_out, activation='sigmoid'))
23  model.compile(loss='binary_crossentropy', optimizer='adam',
    ↪   metrics=['accuracy'])
24
25  print(model.summary())
26
27  # fit the network
28  history = model.fit(seq_array, label_array, epochs=100,
    ↪   batch_size=200, validation_split=0.05, verbose=2,
29           callbacks =
             ↪   [keras.callbacks.EarlyStopping(monitor=
             ↪   'val_loss', min_delta=0, patience=10,
             ↪   verbose=0, mode='min'),
30                       keras.callbacks.ModelCheckpoint
                         ↪   (model_path,monitor='val_loss',
                         ↪   save_best_only=True, mode='min',
                         ↪   verbose=0)]
31           )
32
33  # list all data in history
34  print(history.history.keys())
```

7. Compute and print the results obtained on train and test data sets.

```
1   # summarize history for Accuracy
2   fig_acc = plt.figure(figsize=(10, 10))
3   plt.plot(history.history['accuracy'])
4   plt.plot(history.history['val_accuracy'])
5   plt.title('model accuracy')
6   plt.ylabel('accuracy')
7   plt.xlabel('epoch')
8   plt.legend(['train', 'test'], loc='upper left')
9   plt.show()
10  fig_acc.savefig("model_accuracy.png")
11
12  # summarize history for Loss
13  fig_acc = plt.figure(figsize=(10, 10))
14  plt.plot(history.history['loss'])
15  plt.plot(history.history['val_loss'])
```

```
16  plt.title('model loss')
17  plt.ylabel('loss')
18  plt.xlabel('epoch')
19  plt.legend(['train', 'test'], loc='upper left')
20  plt.show()
21  fig_acc.savefig("model_loss.png")
22
23  # training metrics
24  scores = model.evaluate(seq_array, label_array, verbose=1,
     ↪  batch_size=200)
25  print('Accurracy: {}'.format(scores[1]))
26
27  # make predictions and compute confusion matrix
28  y_pred = model.predict_classes(seq_array,verbose=1,
     ↪  batch_size=200)
29  y_true = label_array
30
31  test_set = pd.DataFrame(y_pred)
32  test_set.to_csv('binary_submit_train.csv', index = None)
33
34  print('Confusion matrix\n- x-axis is true labels.\n- y-axis
     ↪  is predicted labels')
35  cm = confusion_matrix(y_true, y_pred)
36  print(cm)
37
38  # compute precision and recall
39  precision = precision_score(y_true, y_pred)
40  recall = recall_score(y_true, y_pred)
41  print( 'precision = ', precision, '\n', 'recall = ',
     ↪  recall)
```

8. Evaluate the Model on the Validation set. Compute and print the results.

```
1  # We pick the last sequence for each id in the test data
2
3  seq_array_test_last =
     ↪  [test_df[test_df['id']==id][sequence_cols].
     ↪  values[-sequence_length:]
4                    for id in test_df['id'].unique() if
                        ↪  len(test_df[test_df['id']==id])
                        ↪  >= sequence_length]
5
6  seq_array_test_last =
     ↪  np.asarray(seq_array_test_last).astype(np.float32)
7  #print("seq_array_test_last")
```

```
 8  #print(seq_array_test_last)
 9  #print(seq_array_test_last.shape)
10
11  # Similarly, we pick the labels
12
13  #print("y_mask")
14  # serve per prendere solo le label delle sequenze che sono
    ↪    almeno lunghe 50
15  y_mask = [len(test_df[test_df['id']==id]) >=
    ↪    sequence_length for id in test_df['id'].unique()]
16  #print("y_mask")
17  #print(y_mask)
18  label_array_test_last =
    ↪    test_df.groupby('id')['label1'].nth(-1)[y_mask].values
19  label_array_test_last =
    ↪    label_array_test_last.reshape(label_array_test_last.
    ↪    shape[0],1).astype(np.float32)
20  #print(label_array_test_last.shape)
21  #print("label_array_test_last")
22  #print(label_array_test_last)
23
24  # if best iteration's model was saved then load and use it
25  if os.path.isfile(model_path):
26      estimator = load_model(model_path)
27
28  # test metrics
29  scores_test = estimator.evaluate(seq_array_test_last,
    ↪    label_array_test_last, verbose=2)
30  print('Accurracy: {}'.format(scores_test[1]))
31
32  # make predictions and compute confusion matrix
33  y_pred_test =
    ↪    estimator.predict_classes(seq_array_test_last)
34  y_true_test = label_array_test_last
35
36  test_set = pd.DataFrame(y_pred_test)
37  test_set.to_csv('binary_submit_test.csv', index = None)
38
39  print('Confusion matrix\n- x-axis is true labels.\n- y-axis
    ↪    is predicted labels')
40  cm = confusion_matrix(y_true_test, y_pred_test)
41  print(cm)
42
43  # creating a confusion matrix
44  cm = confusion_matrix(y_true_test, y_pred_test)
```

```
45  import pandas as pd
46  import seaborn as sn
47  import matplotlib.pyplot as plt
48  %matplotlib inline
49  df_cm = pd.DataFrame(cm, index = [i for i in "01"],
50                     columns = [i for i in "01"])
51  plt.figure(figsize = (10,7))
52  sn.heatmap(df_cm, annot=True)
53
54
55  # compute precision and recall
56  precision_test = precision_score(y_true_test, y_pred_test)
57  recall_test = recall_score(y_true_test, y_pred_test)
58  f1_test = 2 * (precision_test * recall_test) /
    ↪  (precision_test + recall_test)
59  print( 'Precision: ', precision_test, '\n', 'Recall: ',
    ↪  recall_test,'\n', 'F1-score:', f1_test )
60
61  # Plot in blue color the predicted data and in green color
    ↪  the
62  # actual data to verify visually the accuracy of the model.
63  fig_verify = plt.figure(figsize=(10, 5))
64  plt.plot(y_pred_test, color="blue")
65  plt.plot(y_true_test, color="green")
66  plt.title('prediction')
67  plt.ylabel('value')
68  plt.xlabel('row')
69  plt.legend(['predicted', 'actual data'], loc='upper left')
70  plt.show()
71  fig_verify.savefig("model_verify.png")
```

Figures 7.25 and 7.26 show the results in terms of model accuracy and loss, the confusion matrix for train vs test data sets and the prediction on the validation data set, respectively. The model gives an accuracy of 97.8% on the test data set and 96.7% on the validation data set. The accuracy measures the ratio of the correctly predicted observation to the total observations. In terms of precision, the model achieves 92% precision which represents the ratio of the correctly predicted positive observations to the total predicted positive observations. The recall is 96% representing the ratio of correctly predicted positive observations to the all observations in actual class. While the F1 score obtained is 94% representing the weighted average of Precision and Recall. Meaning that it takes the score of both false positives and false negatives into account.

Regression – which predicts how many cycles an in-service engine still has before it will fail. We use the same LSTM model to perform regression as

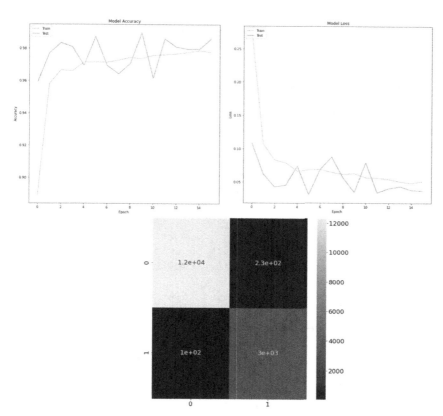

FIGURE 7.25
Binary Classification Train vs. Test Results

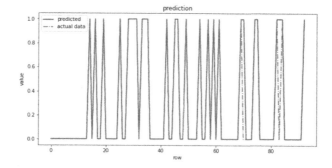

FIGURE 7.26
Binary Classification Validation Results

well. Thus, the initial steps are the same as in the case of binary classification. The changes appear from the fifth step onward. Thus, these steps will be represented below. This time, RUL will be used as the target instead of the binary label.

1. Creating and train the LSTM model. Use the R^2 as the metrics for training

```
 1  def r2_keras(y_true, y_pred):
 2      """Coefficient of Determination
 3      """
 4      SS_res =  K.sum(K.square( y_true - y_pred ))
 5      SS_tot = K.sum(K.square( y_true - K.mean(y_true) ) )
 6      return ( 1 - SS_res/(SS_tot + K.epsilon()) )
 7
 8  # Next, we build a deep network.
 9  # The first layer is an LSTM layer with 100 units followed
    ↪  by another LSTM layer with 50 units.
10  # Dropout is also applied after each LSTM layer to control
    ↪  overfitting.
11  # Final layer is a Dense output layer with single unit and
    ↪  linear activation since this is a regression problem.
12  nb_features = seq_array.shape[2]
13  nb_out = label_array.shape[1]
14
15  model = Sequential()
16  model.add(LSTM(
17          input_shape=(sequence_length, nb_features),
18          units=100,
19          return_sequences=True))
20  model.add(Dropout(0.2))
21  model.add(LSTM(
22          units=50,
23          return_sequences=False))
24  model.add(Dropout(0.2))
25  model.add(Dense(units=nb_out))
26  model.add(Activation("linear"))
27  model.compile(loss='mean_squared_error',
    ↪  optimizer='rmsprop',metrics=['mae',r2_keras])
28
29  print(model.summary())
30
31  # fit the network
32  history = model.fit(seq_array, label_array, epochs=100,
    ↪  batch_size=200, validation_split=0.05, verbose=2,
```

```
33    callbacks = [keras.callbacks.EarlyStopping
      ↪  (monitor='val_loss', min_delta=0,
      ↪  patience=10, verbose=0, mode='min'),
34                keras.callbacks.ModelCheckpoint
                  ↪  (model_path,monitor='val_loss',
                  ↪  save_best_only=True, mode='min',
                  ↪  verbose=0)]
35    )
36
37  # list all data in history
38  print(history.history.keys())
```

2. Compute and print the results obtained on train and test data sets.

```
1   # summarize history for R^2
2   plt.rcParams['font.size'] = 18
3   fig_acc = plt.figure(figsize=(10, 10))
4   plt.plot(history.history['r2_keras'])
5   plt.plot(history.history['val_r2_keras'])
6   plt.title('Model R^2')
7   plt.ylabel('R^2')
8   plt.xlabel('Epoch')
9   plt.legend(['Train', 'Test'], loc='upper left')
10  plt.show()
11  fig_acc.savefig("model_r2.png")
12
13  # summarize history for MAE
14  plt.rcParams['font.size'] = 18
15  fig_acc = plt.figure(figsize=(10, 10))
16  plt.plot(history.history['mae'])
17  plt.plot(history.history['val_mae'])
18  plt.title('Model MAE')
19  plt.ylabel('MAE')
20  plt.xlabel('Epoch')
21  plt.legend(['Train', 'Test'], loc='upper left')
22  plt.show()
23  fig_acc.savefig("model_mae.png")
24
25  # summarize history for Loss
26  plt.rcParams['font.size'] = 18
27  fig_acc = plt.figure(figsize=(10, 10))
28  plt.plot(history.history['loss'])
29  plt.plot(history.history['val_loss'])
30  plt.title('Model Loss')
31  plt.ylabel('Loss')
```

```
32  plt.xlabel('Epoch')
33  plt.legend(['Train', 'Test'], loc='upper left')
34  plt.show()
35  fig_acc.savefig("model_regression_loss.png")
36
37  from sklearn import metrics
38  print('Confusion matrix\n- x-axis is true labels.\n- y-axis
    ↪  is predicted labels')
39  cm = metrics.confusion_matrix(y_true, y_pred)
40  print(cm)
41
42  # creating a confusion matrix
43  cm = confusion_matrix(y_true, y_pred)
44  import pandas as pd
45  import seaborn as sn
46  import matplotlib.pyplot as plt
47  %matplotlib inline
48  df_cm = pd.DataFrame(cm, index = [i for i in "01"],
49                      columns = [i for i in "01"])
50  plt.rcParams['font.size'] = 18
51  plt.figure(figsize = (10,10))
52  sn.heatmap(df_cm, annot=True)
53
54  # training metrics
55  scores = model.evaluate(seq_array, label_array, verbose=1,
    ↪  batch_size=200)
56  print('\nMAE: {}'.format(scores[1]))
57  print('\nR^2: {}'.format(scores[2]))
58
59  y_pred = model.predict(seq_array,verbose=1, batch_size=200)
60  y_true = label_array
61
62  test_set = pd.DataFrame(y_pred)
63  test_set.to_csv('submit_train.csv', index = None)
```

3. Evaluate the Model on the Validation set. Compute and print the results.

```
1  # We pick the last sequence for each id in the test data
2  seq_array_test_last =
   ↪  [test_df[test_df['id']==id][sequence_cols].
   ↪  values[-sequence_length:]
3                      for id in test_df['id'].unique() if
                       ↪  len(test_df[test_df['id']==id])
                       ↪  >= sequence_length]
4
```

```
5   seq_array_test_last =
    ↪   np.asarray(seq_array_test_last).astype(np.float32)
6   #print("seq_array_test_last")
7   #print(seq_array_test_last)
8   #print(seq_array_test_last.shape)
9
10  # Similarly, we pick the labels
11  #print("y_mask")
12  y_mask = [len(test_df[test_df['id']==id]) >=
    ↪   sequence_length for id in test_df['id'].unique()]
13  label_array_test_last =
    ↪   test_df.groupby('id')['RUL'].nth(-1)[y_mask].values
14  label_array_test_last =
    ↪   label_array_test_last.reshape(label_array_test_last.
    ↪   shape[0],1).astype(np.float32)
15  #print(label_array_test_last.shape)
16  #print("label_array_test_last")
17  #print(label_array_test_last)
18
19  # if best iteration's model was saved then load and use it
20  if os.path.isfile(model_path):
21      estimator =
        ↪   load_model(model_path,custom_objects={'r2_keras':
        ↪   r2_keras})
22
23      # test metrics
24      scores_test = estimator.evaluate(seq_array_test_last,
        ↪   label_array_test_last, verbose=2)
25      print('\nMAE: {}'.format(scores_test[1]))
26      print('\nR^2: {}'.format(scores_test[2]))
27
28      y_pred_test = estimator.predict(seq_array_test_last)
29      y_true_test = label_array_test_last
30
31      test_set = pd.DataFrame(y_pred_test)
32      test_set.to_csv('submit_test.csv', index = None)
33
34      # Plot in blue color the predicted data and in green
        ↪   color the
35      # actual data to verify visually the accuracy of the
        ↪   model.
36      plt.rcParams['font.size'] = 18
37      fig_verify = plt.figure(figsize=(10, 5))
38      plt.plot(y_pred_test, color="blue")
39      plt.plot(y_true_test, color="green")
```

```
40    plt.title('prediction')
41    plt.ylabel('value')
42    plt.xlabel('row')
43    plt.legend(['predicted', 'actual data'], loc='upper
      ↪  left')
44    plt.show()
45    fig_verify.savefig("model_regression_verify.png")
```

Figures 7.27 and 7.28 show the results in terms of model R^2, Mean Absolute Error (MAE) loss, the confusion matrix for train vs test data sets and the prediction on the validation data set, respectively. The model returns an MAE value of 16, R^2 value of 0.75 on the test data set and an MAE value of 12 and R^2 value of 0.79 on the validation data set. R^2 value represents the squared correlation between the observed outcome values and the predicted values by the model. Thus the R^2 is a good indicator of how well the model fits the data and generally the higher the value, the better it fits the data, thus indicating a better model. While MAE measures the prediction error regardless if the error is a positive or a negative error.

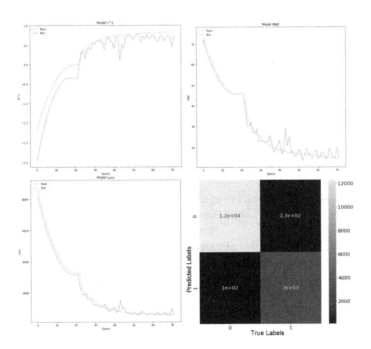

FIGURE 7.27
Regression Train vs. Test Results

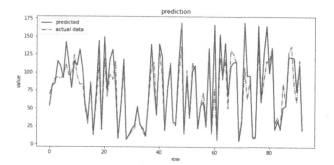

FIGURE 7.28
Regression Validation Results

7.8 Practical Use-Case Scenario: ML for Smart Cities IoT Using Python, Tensorflow, Jupyter

This practical use-case scenario is based on Miniconda[9] release of Python which already includes most of the required packages. At the time of the writing this use-case scenario was run using Python 3.6.10, TensorFlow Version 2.3, Keras Version 2.4.0 and Jupyter Notebook version 6.1.1.

The aim of this use-case scenario is to see how we can use ML for detecting crime using San Francisco crime data [45]. The code for *San Francisco Crime Data* using different ML solutions, has been implemented by Susan Li [81] and is available on GitHub.[10]

7.8.1 Use-Case Scenario Settings

For the purpose of this use-case scenario we are going to look into detecting crime using San Francisco crime data, available online on the city's open data portal.[11]

We are interested in a system that could classify a crime description into different categories and that could automatically assign a described crime to a category which could help law enforcement to delegate the right officers to the crime. Additionally, the system could automatically assign officers to the crime based on the classification. Consequently, the main objective of this use-case scenario is to train a model based on 39 pre-defined categories and test the model accuracy using different ML algorithms. For any new crime

---

[9]https://docs.conda.io/en/latest/miniconda.html#windows-installers
[10]https://github.com/susanli2016/Machine-Learning-with-Python/blob/master/SF_Crime_Text_Classification_PySpark.ipynb
[11]https://data.sfgov.org/Public-Safety/Police-Department-Incident-Reports-Historical-2003/tmnf-yvry

description, the system should be able to assign it to one of 39 categories. This is multi-class text classification problem, and it is assumed that each new crime description is assigned to one and only one category. We will first use a variety of feature extraction technique along with different supervised machine learning algorithms in Spark.

7.8.2   Data Set Summary

The dataset used in this lab consists of the following main attributes [82]:

- IncidntNum: It is a numerical field. Denotes the incident number of the crime as recorded in the police logs. It is analogous to the row number.

- Descript: It is a text field. Contains a brief description about the crime. This field provides slightly more information than the Category field but is still quite limited.

- DayOfWeek: It is a text field. Specifies the day of the week when the crime occurred. It takes on one of the values from: Monday, Tuesday, Wednesday, Thursday, Friday, Saturday, Sunday.

- Date: It is a Date-Time field. Specifies the exact date of the crime.

- PdDistrict: It is a text field. Specifies the police district the crime occurred in. San Francisco has been divided in 10 police districts. It takes on one of the values from: Southern, Tenderloin, Mission, Central, Northern, Bayview, Richmond, Taraval, Ingleside, Park.

- Resolution: It is a text field. Specifies the resolution for the crime. It takes one of these values: Arrested, Booked, None.

- Address: It is a text field. Gives the street address of the crime.

- X: It is a geographic field. It gives the longitudinal coordinates of the crime.

- Y: It is a geographic field. It gives the latitudinal coordinates of the crime.

- Location: It is a location field. It is in the form of a pair of coordinates, i.e. (X, Y).

- PdId: It is a numerical field. It is a unique identifier for each complaint registered. It is used in the database update or search operations.

- Category: It is a text field. Specifies the category of the crime. Originally, there are 39 distinct values (such as Assault, Larceny/Theft, Prostitution, etc.) in this field. It is also the dependent variable we will try to predict for the test set.

There are about 1.4 million rows in the dataset and the size of the dataset is approximately 500 MB. It contains data from the year 2003 to (May) 2018. The data set contains several columns, however in this use-case scenario we only need Category and Descript fields for training and testing datasets. A snapshot of these two categories is illustrated in Figure 7.29.

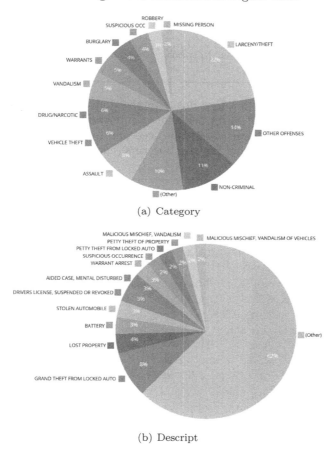

(a) Category

(b) Descript

FIGURE 7.29
San Francisco Dataset Category and Descript Fields

7.8.3 Source Code and Results

This subsection explains the main building source code blocks used to solve this problem using the several approaches.

1. Data Ingestion and Extraction – loading the data set file using Spark csv packages and explore the data.

```
1  from pyspark.sql import SQLContext
2  from pyspark import SparkContext
3  sc =SparkContext()
4  sqlContext = SQLContext(sc)
5
6  data = sqlContext.read.format('com.databricks.spark.csv').
   ↪  options(header='true',
   ↪  inferschema='true').load('sf_crime_dataset.csv')
7
```

2. As mentioned, the data contains several columns, however we will only
   need Category and Descript fields for training and testing data sets. Thus
   we removed the columns we do not need and have a look at the first five
   rows. We can also print the top 20 crime categories and descriptors as
   well.

```
1  drop_list = ['PdId', 'IncidntNum', 'Incident Code',
   ↪  'DayOfWeek', 'Date', 'Time', 'PdDistrict',
   ↪  'Resolution', 'Address', 'X', 'Y', 'location',
2  'SF Find Neighborhoods 2 2', 'Current Police Districts 2
   ↪  2', 'Current Supervisor Districts 2 2', 'Analysis
   ↪  Neighborhoods 2 2', 'DELETE - Fire Prevention
   ↪  Districts 2 2', 'DELETE - Police Districts 2 2',
   ↪  'DELETE - Supervisor Districts 2 2', 'DELETE - Zip
   ↪  Codes 2 2', 'DELETE - Neighborhoods 2 2', 'DELETE -
   ↪  2017 Fix It Zones 2 2', 'Civic Center Harm Reduction
   ↪  Project Boundary 2 2', 'Fix It Zones as of 2017-11-06
   ↪  2 2', 'DELETE - HSOC Zones 2 2', 'Fix It Zones as of
   ↪  2018-02-07 2 2', 'CBD, BID and GBD Boundaries as of
   ↪  2017 2 2', 'Areas of Vulnerability, 2016 2 2',
   ↪  'Central Market/Tenderloin Boundary 2 2', 'Central
   ↪  Market/Tenderloin Boundary Polygon - Updated 2 2',
   ↪  'HSOC Zones as of 2018-06-05 2 2', 'OWED Public Spaces
   ↪  2 2', 'Neighborhoods 2']
3
4  data = data.select([column for column in data.columns if
   ↪  column not in drop_list])
5  data.show(5)
6
7  from pyspark.sql.functions import col
8
9  # by top 20 categories
10 data.groupBy("Category") \
11     .count() \
```

```
12    .orderBy(col("count").desc()) \
13    .show()
14
15  # by top 20 descriptions
16  data.groupBy("Descript") \
17    .count() \
18    .orderBy(col("count").desc()) \
19    .show()
20
```

3. Model Pipeline – We need to perform text processing as the data set has textual data. Thus, the pipeline includes three steps:

- regexTokenizer: Tokenization of the sentence into a list of words
- stopwordsRemover: Remove Sandard Stop Words like: "http", "https", "amp", "rt", "t", "c", "the"
- countVectors: convert the words to numeric vector (features) used as input to train the model.

```
1   from pyspark.ml.feature import RegexTokenizer,
     ↪   StopWordsRemover, CountVectorizer
2   from pyspark.ml.classification import LogisticRegression
3
4   # regular expression tokenizer
5   regexTokenizer = RegexTokenizer(inputCol="Descript",
     ↪   outputCol="words", pattern="\\W")
6
7   # stop words
8   add_stopwords = ["http","https","amp","rt","t","c","the"] #
     ↪   standard stop words
9
10  stopwordsRemover = StopWordsRemover(inputCol="words",
     ↪   outputCol="filtered").setStopWords(add_stopwords)
11
12  # bag of words count
13  countVectors = CountVectorizer(inputCol="filtered",
     ↪   outputCol="features", vocabSize=10000, minDF=5)
```

4. StringIndexer encodes a string column of labels to a column of label indices. The indices are in [0, numLabels), ordered by label frequencies, so the most frequent label gets index 0. In our case, the label column (Category) will be encoded to label indices, from 0 to 38; the most frequent label (LARCENY/THEFT) will be indexed as 0.

```
1  from pyspark.ml.feature import OneHotEncoder,
   ↪  StringIndexer, VectorAssembler
2  label_stringIdx = StringIndexer(inputCol = "Category",
   ↪  outputCol = "label")
3
4  from pyspark.ml import Pipeline
5
6  pipeline = Pipeline(stages=[regexTokenizer,
   ↪  stopwordsRemover, countVectors, label_stringIdx])
7
8  # Fit the pipeline to training documents.
9  pipelineFit = pipeline.fit(data)
10 dataset = pipelineFit.transform(data)
```

5. Partition Training & Test sets – the data set is split into training and test data sets.

```
1  ### Randomly split data into training and test sets. set
   ↪  seed for reproducibility
2  (trainingData, testData) = dataset.randomSplit([0.7, 0.3],
   ↪  seed = 100)
3  print("Training Dataset Count: " +
   ↪  str(trainingData.count()))
4  print("Test Dataset Count: " + str(testData.count()))
```

6. **Logistic Regression using Count Vector Features** – the model will make predictions and score on the test set; the top 10 predictions from the highest probability are listed.

```
1  # Build the model
2  lr = LogisticRegression(maxIter=20, regParam=0.3,
   ↪  elasticNetParam=0)
3
4  # Train model with Training Data
5  lrModel = lr.fit(trainingData)
6  predictions = lrModel.transform(testData)
7
8  predictions.filter(predictions['prediction'] == 0) \
9      .select("Descript","Category","probability",
   ↪  "label","prediction") \
10     .orderBy("probability", ascending=False) \
11     .show(n = 10, truncate = 30)
12 from pyspark.ml.evaluation import
   ↪  MulticlassClassificationEvaluator
```

```
13  evaluator = MulticlassClassificationEvaluator(predictionCol
    ↪ ="prediction")
14  evaluator.evaluate(predictions)
15
```

7. **Output:** On the test dataset, the model provides a 97% accuracy.

8. **Logistic Regression using TF-IDF Features**

```
1   from pyspark.ml.feature import HashingTF, IDF
2
3   # Add HashingTF and IDF to transformation
4   hashingTF = HashingTF(inputCol="filtered",
    ↪ outputCol="rawFeatures", numFeatures=10000)
5   idf = IDF(inputCol="rawFeatures", outputCol="features",
    ↪ minDocFreq=5) #minDocFreq: remove sparse terms
6
7   # Redo Pipeline
8   pipeline = Pipeline(stages=[regexTokenizer,
    ↪ stopwordsRemover, hashingTF, idf, label_stringIdx])
9
10  pipelineFit = pipeline.fit(data)
11  dataset = pipelineFit.transform(data)
12
13  ### Randomly split data into training and test sets. set
    ↪ seed for reproducibility
14  (trainingData, testData) = dataset.randomSplit([0.7, 0.3],
    ↪ seed = 100)
15
16  # Build the model
17  lr = LogisticRegression(maxIter=20, regParam=0.3,
    ↪ elasticNetParam=0)
18
19  # Train model with Training Data
20  lrModel = lr.fit(trainingData)
21
22  predictions = lrModel.transform(testData)
23
24  predictions.filter(predictions['prediction'] == 0) \
25      .select("Descript","Category","probability",
        ↪ "label","prediction") \
26      .orderBy("probability", ascending=False) \
27      .show(n = 10, truncate = 30)
28
```

```
29    evaluator =
      ↪  MulticlassClassificationEvaluator(predictionCol=
      ↪  "prediction")
30  evaluator.evaluate(predictions)
```

9. **Output:** On the test dataset, the model provides a 97% accuracy.

10. **Cross Validation** – try cross-validation to tune the hyper parameters, and only tune the count vectors Logistic Regression.

```
1   pipeline = Pipeline(stages=[regexTokenizer,
    ↪  stopwordsRemover, countVectors, label_stringIdx])

2
3   pipelineFit = pipeline.fit(data)
4   dataset = pipelineFit.transform(data)
5   (trainingData, testData) = dataset.randomSplit([0.7, 0.3],
    ↪  seed = 100)

6
7   # Build the model
8   lr = LogisticRegression(maxIter=20, regParam=0.3,
    ↪  elasticNetParam=0)

9
10  from pyspark.ml.tuning import ParamGridBuilder,
    ↪  CrossValidator

11
12  # Create ParamGrid for Cross Validation
13  paramGrid = (ParamGridBuilder()
14               .addGrid(lr.regParam, [0.1, 0.3, 0.5]) #
                 ↪  regularization parameter
15               .addGrid(lr.elasticNetParam, [0.0, 0.1, 0.2])
                 ↪  # Elastic Net Parameter (Ridge = 0)
16  #            .addGrid(model.maxIter, [10, 20, 50]) #Number
    ↪  of iterations
17  #            .addGrid(idf.numFeatures, [10, 100, 1000]) #
    ↪  Number of features
18               .build())
19
20  # Create 5-fold CrossValidator
21  cv = CrossValidator(estimator=lr, \
22                      estimatorParamMaps=paramGrid, \
23                      evaluator=evaluator, \
24                      numFolds=5)
25
26  # Run cross validations
27  cvModel = cv.fit(trainingData)
```

```
28  # this will likely take a fair amount of time because of
    ↪    the amount of models that we're creating and testing
29
30  # Use test set here so we can measure the accuracy of our
    ↪    model on new data
31  predictions = cvModel.transform(testData)
32
33  # cvModel uses the best model found from the Cross
    ↪    Validation
34  # Evaluate best model
35  evaluator =
    ↪    MulticlassClassificationEvaluator(predictionCol=
    ↪    "prediction")
36  evaluator.evaluate(predictions)
37
```

11. **Output:** On the test dataset, the model provides a 99% accuracy.

12. **Naive Bayes**

```
1   from pyspark.ml.classification import NaiveBayes
2
3   # create the trainer and set its parameters
4   nb = NaiveBayes(smoothing=1)
5
6   # train the model
7   model = nb.fit(trainingData)
8   predictions = model.transform(testData)
9   predictions.filter(predictions['prediction'] == 0) \
10      .select("Descript","Category","probability",
        ↪    "label","prediction") \
11      .orderBy("probability", ascending=False) \
12      .show(n = 10, truncate = 30)
13  evaluator =
    ↪    MulticlassClassificationEvaluator(predictionCol=
    ↪    "prediction")
14  evaluator.evaluate(predictions)
15
```

13. **Output:** On the test dataset, the model provides a 99% accuracy.

14. **Random Forest**

```
1   from pyspark.ml.classification import
    ↪    RandomForestClassifier
2
```

```
3    # Create an initial RandomForest model.
4    rf = RandomForestClassifier(labelCol="label", \
5                                featuresCol="features", \
6                                numTrees = 100, \
7                                maxDepth = 4, \
8                                maxBins = 32)
9
10   # Train model with Training Data
11   rfModel = rf.fit(trainingData)
12   predictions = rfModel.transform(testData)
13
14   predictions.filter(predictions['prediction'] == 0) \
15       .select("Descript","Category","probability",
            ↪   "label","prediction") \
16       .orderBy("probability", ascending=False) \
17       .show(n = 10, truncate = 30)
18
19   evaluator =
        ↪   MulticlassClassificationEvaluator(predictionCol=
        ↪   "prediction")
20   evaluator.evaluate(predictions)
21
```

15. **Output:** On the test dataset, the model provides a 72% accuracy.

Looking at the results we can notice that Random Forest performs very well, representing robust and versatile method. However, when we deal with high-dimensional sparse data the Random Forest model is not the best choice. Consequently, the obvious choice for this use-case scenario is the Logistic Regression with cross-validation.

Bibliography

[1] Mehmet Fatih Tuysuz and Ramona Trestian. From serendipity to sustainable green iot: Technical, industrial and political perspective. *Computer Networks*, 182:107469, 2020.

[2] Innovation Labs. Lecture 4: Introduction to iot. https://ocw.cs.pub.ro/courses/iot2016/courses/04. Accessed: 2022-05-27.

[3] Scott Goldfine. 5g is coming, here's how it will effect iot and the security industry. https://www.securitysales.com/columns/5g-effect-iot-security-industry/. Accessed: 2022-05-27.

[4] Qualcomm announces the launch of healthycircles mobile. `https://www.qualcomm.com/news/releases/2015/04/08`. Accessed: 2022-05-27.

[5] And finally: Nike to debut connected shoes, but don't expect them to track your run. `https://www.wareable.com/wearable-tech/nike-connected-shoe-2801`. Accessed: 2022-05-27.

[6] C. E. Anagnostopoulos, I. E. Anagnostopoulos, I. D. Psoroulas, V. Loumos, and E. Kayafas. License plate recognition from still images and video sequences: A survey. *IEEE Transactions on Intelligent Transportation Systems*, 9(3):377–391, Sept 2008.

[7] S. He, Y. Yuan, C. Fu, X. Hu, and Y. Zhao. Robust license plate detection using profile-based filter. In *2018 Tenth International Conference on Advanced Computational Intelligence (ICACI)*, pages 794–800, March 2018.

[8] A. M. Al-Ghaili, S. Mashohor, A. R. Ramli, and A. Ismail. Vertical-edge-based car-license-plate detection method. *IEEE Transactions on Vehicular Technology*, 62(1):26–38, Jan 2013.

[9] Jun-Wei Hsieh, Shih-Hao Yu, and Yung-Sheng Chen. Morphology-based license plate detection from complex scenes. In *Object recognition supported by user interaction for service robots*, volume 3, pages 176–179 vol.3, Aug 2002.

[10] A. A. Lensky, K. Jo, and V. V. Gubarev. Vehicle license plate detection using local fractal dimension and morphological analysis. In *2006 International Forum on Strategic Technology*, pages 47–50, Oct 2006.

[11] L. Luo, H. Sun, W. Zhou, and L. Luo. An efficient method of license plate location. In *2009 First International Conference on Information Science and Engineering*, pages 770–773, Dec 2009.

[12] L. Hu and Q. Ni. Iot-driven automated object detection algorithm for urban surveillance systems in smart cities. *IEEE Internet of Things Journal*, 5(2):747–754, April 2018.

[13] O. Bulan, V. Kozitsky, P. Ramesh, and M. Shreve. Segmentation- and annotation-free license plate recognition with deep localization and failure identification. *IEEE Transactions on Intelligent Transportation Systems*, 18(9):2351–2363, Sept 2017.

[14] Muhammad Aasim Rafique, Witold Pedrycz, and Moongu Jeon. Vehicle license plate detection using region-based convolutional neural networks. *Soft Computing*, pages 1–12, June 2017.

[15] T. Juhana and V. G. Anggraini. Design and implementation of smart home surveillance system. In *2016 10th International Conference on*

Telecommunication Systems Services and Applications (TSSA), pages 1–5, Oct 2016.

[16] L. Chen, Chi-Ren Chen, and Da-En Chen. Vips: A video-based indoor positioning system with centimeter-grade accuracy for the iot. In *2017 IEEE International Conference on Pervasive Computing and Communications Workshops (PerCom Workshops)*, pages 63–65, March 2017.

[17] M. Alam, J. Ferreira, S. Mumtaz, M. A. Jan, R. Rebelo, and J. A. Fonseca. Smart cameras are making our beaches safer: A 5g-envisioned distributed architecture for safe, connected coastal areas. *IEEE Vehicular Technology Magazine*, 12(4):50–59, Dec 2017.

[18] A. Giyenko and Y. I. Cho. Intelligent uav in smart cities using iot. In *2016 16th International Conference on Control, Automation and Systems (ICCAS)*, pages 207–210, Oct 2016.

[19] P. Porambage, J. Okwuibe, M. Liyanage, M. Ylianttila, and T. Taleb. Survey on multi-access edge computing for internet of things realization. *IEEE Communications Surveys Tutorials*, pages 1–1, 2018.

[20] Heiko G. Seif and Xiaolong Hu. Autonomous Driving in the iCity—HD Maps as a Key Challenge of the Automotive Industry. *Engineering*, 2(2):159–162, 2016.

[21] J. Jiao. Machine learning assisted high-definition map creation. In *2018 IEEE 42nd Annual Computer Software and Applications Conference (COMPSAC)*, volume 01, pages 367–373, July 2018.

[22] J. J. P. C. Rodrigues, D. B. De Rezende Segundo, H. A. Junqueira, M. H. Sabino, R. M. Prince, J. Al-Muhtadi, and V. H. C. De Albuquerque. Enabling technologies for the internet of health things. *IEEE Access*, 6:13129–13141, 2018.

[23] C. Doukas and I. Maglogiannis. Bringing iot and cloud computing towards pervasive healthcare. In *2012 Sixth International Conference on Innovative Mobile and Internet Services in Ubiquitous Computing*, pages 922–926, July 2012.

[24] A. Dohr, R. Modre-Opsrian, M. Drobics, D. Hayn, and G. Schreier. The internet of things for ambient assisted living. In *Proceedings of the 2010 Seventh International Conference on Information Technology: New Generations*, ITNG '10, pages 804–809, Washington, DC, USA, 2010. IEEE Computer Society.

[25] J. Wu, D. Chou, and J. Jiang. The virtual environment of things (veot): A framework for integrating smart things into networked virtual environments. In *IEEE International Conference on Internet of Things (iThings), and Green Computing and Communications (GreenCom) and*

Cyber, Physical and Social Computing (CPSCom), pages 456–459, Sept 2014.

[26] M. Alessi, E. Giangreco, M. Pinnella, S. Pino, D. Storelli, L. Mainetti, V. Mighali, and L. Patrono. A web based virtual environment as a connection platform between people and IoT. *2016 International Multidisciplinary Conference on Computer and Energy Science, SpliTech 2016*, 2016.

[27] I. Toumpalidis, K. Cheliotis, F. Roumpani, and A. Hudson-Smith. Vr binoculars: An immersive visualization framework for iot data streams. In *Living in the Internet of Things: Cybersecurity of the IoT – 2018*, pages 1–7, March 2018.

[28] Myeong In Choi, Lee Won Park, Sanghoon Lee, Jun Yeon Hwang, and Sehyun Park. Design and implementation of Hyper-connected IoT-VR Platform for customizable and intuitive remote services. *2017 IEEE International Conference on Consumer Electronics, ICCE 2017*, pages 396–397, 2017.

[29] Anderson Augusto Simiscuka and Gabriel-Miro Muntean. Synchronisation between Real and Virtual-World Devices in a VR-IoT Environment. *IEEE International Symposium on Broadband Multimedia Systems*, 2018.

[30] Zhihan Lv, Tengfei Yin, Houbing Song, and Ge Chen. Virtual Reality Smart City Based on WebVRGIS. *IEEE Internet of Things Journal*, 4662(c):1–1, 2016.

[31] J. Westlin and T. H. Laine. Short paper: Calory battle ar: An extensible mobile augmented reality exergame platform. In *2014 IEEE World Forum on Internet of Things (WF-IoT)*, pages 171–172, March 2014.

[32] Amanda E. Staiano and Sandra L. Calvert. Exergames for Physical Education Courses: Physical, Social, and Cognitive Benefits. *Child Development Perspectives*, 5(2):93–98, June 2011.

[33] Muhammad Munwar Iqbal, Muhammad Farhan, Sohail Jabbar, Yasir Saleem, and Shehzad Khalid. Multimedia based IoT-centric smart framework for eLearning paradigm. *Multimedia Tools and Applications*, pages 1–20, February 2018.

[34] S. Creane, Y. Crotty, and M. Farren. A proposed use of virtual and augmented reality for supporting inquiry based learning. In *2015 International Conference on Interactive Mobile Communication Technologies and Learning (IMCL)*, pages 393–395, Nov 2015.

[35] M. S. Mekala and P. Viswanathan. A novel technology for smart agriculture based on iot with cloud computing. In *2017 International Conference on I-SMAC (IoT in Social, Mobile, Analytics and Cloud) (I-SMAC)*, pages 75–82, Feb 2017.

[36] Ji chun Zhao, Jun feng Zhang, Yu Feng, and Jian xin Guo. The study and application of the iot technology in agriculture. In *2010 3rd International Conference on Computer Science and Information Technology*, volume 2, pages 462–465, July 2010.

[37] R. Nukala, K. Panduru, A. Shields, D. Riordan, P. Doody, and J. Walsh. Internet of things: A review from 'farm to fork'. In *2016 27th Irish Signals and Systems Conference (ISSC)*, pages 1–6, June 2016.

[38] Bharat S Chaudhari, Marco Zennaro, and Suresh Borkar. Lpwan technologies: Emerging application characteristics, requirements, and design considerations. *Future Internet*, 12(3):46, 2020.

[39] Difference between lora and lorawan. https://www.choovio.com/difference-between-lora-and-lorawan/. Accessed: 2022-05-27.

[40] Lora link-budget and sensitivity calculations – example explained. https://www.techplayon.com/lora-link-budget-sensitivity-calculations-example-explained/. Accessed: 2022-05-27.

[41] Francesca Cuomo, Manuel Campo, Alberto Caponi, Giuseppe Bianchi, Giampaolo Rossini, and Patrizio Pisani. Explora: Extending the performance of lora by suitable spreading factor allocations. In *2017 IEEE 13th International Conference on Wireless and Mobile Computing, Networking and Communications (WiMob)*, pages 1–8. IEEE, 2017.

[42] Toni Adame, Albert Bel, Boris Bellalta, Jaume Barcelo, and Miquel Oliver. Ieee 802.11 ah: the wifi approach for m2m communications. *IEEE Wireless Communications*, 21(6):144–152, 2014.

[43] Jochen H Schiller. *Mobile communications*. Pearson education, 2003.

[44] Andreas Lindholm, Niklas Wahlström, Fredrik Lindsten, and Thomas B Schön. *Machine Learning: A First Course for Engineers and Scientists*. Cambridge University Press, 2022.

[45] Amita Kapoor. *Hands-On Artificial Intelligence for IoT: Expert machine learning and deep learning techniques for developing smarter IoT systems*. Packt Publishing Ltd, 2019.

[46] Eugenio Pasqua. How 5g, ai and iot enable "intelligent connectivity". https://iot-analytics.com/how-5g-ai-and-iot-enable-intelligent-connectivity/. Accessed: 2022-05-27.

[47] Ethem Alpaydin. *Introduction to machine learning*. MIT press, 2020.

[48] Baotong Chen, Jiafu Wan, Lei Shu, Peng Li, Mithun Mukherjee, and Boxing Yin. Smart factory of industry 4.0: Key technologies, application case, and challenges. *Ieee Access*, 6:6505–6519, 2017.

[49] Qinglin Qi and Fei Tao. Digital twin and big data towards smart manufacturing and industry 4.0: 360 degree comparison. *Ieee Access*, 6:3585–3593, 2018.

[50] Werner Kritzinger, Matthias Karner, Georg Traar, Jan Henjes, and Wilfried Sihn. Digital twin in manufacturing: A categorical literature review and classification. *IFAC-PapersOnLine*, 51(11):1016–1022, 2018.

[51] Jinfeng Liu, Honggen Zhou, Xiaojun Liu, Guizhong Tian, Mingfang Wu, Liping Cao, and Wei Wang. Dynamic evaluation method of machining process planning based on digital twin. *IEEE Access*, 7:19312–19323, 2019.

[52] Brent Daniel Mittelstadt, Patrick Allo, Mariarosaria Taddeo, Sandra Wachter, and Luciano Floridi. The ethics of algorithms: Mapping the debate. *Big Data & Society*, 3(2):2053951716679679, 2016.

[53] Mohsin Raza, Priyan Malarvizhi Kumar, Dang Viet Hung, William Davis, Huan Nguyen, and Ramona Trestian. A digital twin framework for industry 4.0 enabling next-gen manufacturing. In *2020 9th International Conference on Industrial Technology and Management (ICITM)*, pages 73–77. IEEE, 2020.

[54] Stefan Mihai, William Davis, Dang Viet Hung, Ramona Trestian, Mehmet Karamanoglu, Balbir Barn, Raja Prasad, Hrishikesh Venkataraman, and Huan X Nguyen. A digital twin framework for predictive maintenance in industry 4.0. In *Proceedings of the 2020 International Conference on High Performance Computing & Simulation. In: HPCS 2020: 18th Annual Meeting, 22-27 March 2021, Barcelona, Spain (Online Virtual Conference)*, 2021.

[55] William Davis, Mahnoor Yaqoob, Luke Bennett, Stefan Mihai, Dang Viet Hung, Ramona Trestian, Mehmet Karamanoglu, Balbir Barn, and Huan X Nguyen. An innovative blockchain-based traceability framework for industry 4.0 cyber-physical factory. In *11th International Conference on Industrial Technology and Management, 18-20 Feb 2022, Oxford, United Kingdom*, 2022.

[56] Yan Xu, Yanming Sun, Xiaolong Liu, and Yonghua Zheng. A digital-twin-assisted fault diagnosis using deep transfer learning. *IEEE Access*, 7:19990–19999, 2019.

[57] Jinjiang Wang, Lunkuan Ye, Robert X Gao, Chen Li, and Laibin Zhang. Digital twin for rotating machinery fault diagnosis in smart manufacturing. *International Journal of Production Research*, 57(12):3920–3934, 2019.

[58] Roberto Revetria, Flavio Tonelli, Lorenzo Damiani, Melissa Demartini, Federico Bisio, and Nicola Peruzzo. A real-time mechanical structures monitoring system based on digital twin, iot and augmented reality. In *2019 Spring Simulation Conference (SpringSim)*, pages 1–10. IEEE, 2019.

[59] Yizheng Liao, Mark Mollineaux, Richard Hsu, Rebekah Bartlett, Anubhav Singla, Adnan Raja, Ravneet Bajwa, and Ram Rajagopal. Snowfort: An open source wireless sensor network for data analytics in infrastructure and environmental monitoring. *IEEE Sensors Journal*, 14(12):4253–4263, 2014.

[60] Stefan Boschert, Christoph Heinrich, and Roland Rosen. Next generation digital twin. In *Proc. tmce*, volume 2018, pages 7–11. Las Palmas de Gran Canaria, Spain, 2018.

[61] Huan X Nguyen, Ramona Trestian, Duc To, and Mallik Tatipamula. Digital twin for 5g and beyond. *IEEE Communications Magazine*, 59(2):10–15, 2021.

[62] Felix Foo. Digital twins catalyst reflections from digital transformation world. https://www.ericsson.com/en/blog/2019/6/digital-twins-catalyst-booth-reflections-from-digital-transformation-world. Accessed: 2022-05-27.

[63] Huawei launches industry's first site digital twins based 5g digital engineering solution. https://www.huawei.com/en/press-events/news/2020/2/site-digital-twins-based-5g-digital-engineering-solution. Accessed: 2022-05-27.

[64] Ioannis Tomkos, Dimitrios Klonidis, Evangelos Pikasis, and Sergios Theodoridis. Toward the 6g network era: Opportunities and challenges. *IT Professional*, 22(1):34–38, 2020.

[65] Ramona Trestian, Guodong Xie, Pintu Lohar, Edoardo Celeste, Malika Bendechache, Rob Brennan, Evgeniia Jayasekera, Regina Connolly, and Irina Tal. Privacy in a time of covid-19: How concerned are you? *IEEE Security & Privacy*, 19(5):26–35, 2021.

[66] Guodong Xie, Pintu Lohar, Claudia Florea, Malika Bendechache, Ramona Trestian, Rob Brennan, Regina Connolly, and Irina Tal. Privacy in times of covid-19: a pilot study in the republic of ireland. In *The 16th International Conference on Availability, Reliability and Security*, pages 1–6, 2021.

[67] Pintu Lohar, Guodong Xie, Malika Bendechache, Rob Brennan, Edoardo Celeste, Ramona Trestian, and Irina Tal. Irish attitudes toward covid tracker app & privacy: sentiment analysis on twitter and survey data. In *The 16th International Conference on Availability, Reliability and Security*, pages 1–8, 2021.

[68] Pintu Lohar, Guodong Xie, Malika Bendechache, Rob Brennan, Edoardo Celeste, Ramona Trestian, and Irina Tal. Irish attitudes to privacy in covid-19 times: sentiment analysis on twitter and survey data. 2021.

[69] Malika Bendechache, Pintu Lohar, Guodong Xie, Rob Brennan, Ramona Trestian, Edoardo Celeste, Kristina Kapanova, Evgeniia Jayasekera, and Irina Tal. Public attitudes towards privacy in covid-19 times in the republic of ireland: A pilot study. *Information Security Journal: A Global Perspective*, 30(5):281–293, 2021.

[70] Ramona Trestian, Guodong Xie, Pintu Lohar, Edoardo Celeste, Malika Bendechache, Rob Brennan, and Irina Tal. Privatt-a closer look at people's data privacy attitudes in times of covid-19. In *2021 IEEE International Mediterranean Conference on Communications and Networking (MeditCom)*, pages 174–179. IEEE, 2021.

[71] Devin Skoll, Jennifer C Miller, and Leslie A Saxon. Covid-19 testing and infection surveillance: Is a combined digital contact-tracing and mass-testing solution feasible in the united states? *Cardiovascular Digital Health Journal*, 1(3):149–159, 2020.

[72] Nadeem Ahmed, Regio A Michelin, Wanli Xue, Sushmita Ruj, Robert Malaney, Salil S Kanhere, Aruna Seneviratne, Wen Hu, Helge Janicke, and Sanjay K Jha. A survey of covid-19 contact tracing apps. *IEEE Access*, 8:134577–134601, 2020.

[73] Jinfeng Li and Xinyi Guo. Covid-19 contact-tracing apps: A survey on the global deployment and challenges. *arXiv preprint arXiv:2005.03599*, 2020.

[74] Viktoriia Shubina, Sylvia Holcer, Michael Gould, and Elena Simona Lohan. Survey of decentralized solutions with mobile devices for user location tracking, proximity detection, and contact tracing in the Covid-19 era. *Data*, 5(4):87, 2020.

[75] Muhammad Ajmal Azad, Junaid Arshad, Syed Muhammad Ali Akmal, Farhan Riaz, Sidrah Abdullah, Muhammad Imran, and Farhan Ahmad. A first look at privacy analysis of covid-19 contact-tracing mobile applications. *IEEE Internet of Things Journal*, 8(21):15796–15806, 2020.

[76] Dong Wang and Fang Liu. Privacy risk and preservation for covid-19 contact tracing apps. *arXiv preprint arXiv:2006.15433*, 2020.

[77] Documentation for pan-european privacy-preserving proximity tracing (pepp-pt). `https://github.com/pepp-pt/pepp-pt-documentation`. Accessed: 2022-05-27.

[78] Decentralized privacy-preserving proximity tracing. `https://github.com/DP\-3T/documents/blob/master/DP3T\%20White\%20Paper`. Accessed: 2022-05-27.

[79] Exposure notification api launches to support public health agencies. `https://blog.google/inside-google/company-announcements/` `apple-google-exposure-notification-api-launches/`.   Accessed: 2022-05-27.

[80] Umberto Griffo. Predictive maintenance using lstm. 2018, `https://` `github.com/umbertogriffo/Predictive-Maintenance-using-LSTM`.

[81] Susan Li. Detecting crime using san francisco crime data. 2018, `https:` `//github.com/susanli2016/Machine-Learning-with-Python/blob/` `master/SF_Crime_Text_Classification_PySpark.ipynb`.

[82] Isha Pradhan. Exploratory data analysis and crime prediction in san francisco. 2018.

List of Acronyms

AAS	Advanced Antenna System
AMPS	Advanced Mobile Phone Service
AR	Augmented Reality
AI	Artifical Intelligence
ANSI	American National Standards Institute
AM	Amplitude Modulation
ASK	Amplitude Shift Keying
ACI	Adjacent Channel Interference
ARQ	Automatic Repeat Request
AP	Access Point
AHP	Analytic Hierarchy Process
AuC	Authentication Centre
AODV	Ad Hoc On-Demand Distance Vector
AIFS	Arbitration Inter-Frame Spacing
BBU	Baseband Unit
BTS	Base Transceiver Station
BPSK	Binary Phase Shift Keying
BER	Bit Error Rate
BS	Base Station
BSS	Base Station Subsystem
BSC	Base Station Controller
BCI	Brain Computer Interface
BLE	Bluetooth Low Energy

CAV   Connected and Automated Vehicle

C-RAN Cloud Radio Access Network

CDM   Code Division Multiplexing

CDMA Code Division Multiple Access

CDMAone Code Division Multiple Access one

CSMA Carrier Sense Multiple Access

CSMA/CD Carrier Sense Multiple Access with Collision Detection

CSMA/CA Carrier Sense Multiple Access with Collision Avoidance

CTS   Clear to Send

CCI   Co-Channel Interference

CoMP   Coordinated Multi-Point

CapEx Capital expenditures

3CLS   Convergence of Communications, Computing, Control, Localization and Sensing

C-V2X Cellular-V2X

DAMA Demand Assigned Multiple Access

DT   Digital Twin

DU   Digital Unit

DSSS   Direct Sequence Spread Spectrum

DECT   Digital Enhanced Cordless Telecommunications

DPM   Dominant Path Model

DSA   Dynamic Sub-carrier Assignment

DSRC   Dedicated Short-Range Communications

DCF   Distributed Coordination Function

DSDV   Destination-Sequenced Distance Vector

DSR   Dynamic Source Routing

EDGE   Enhanced Data for Global Evolution

eMBB   enhanced Mobile Broadband

eURLLC enhanced Ultra-Reliable and Low Latency Communication

ETSI European Telecommunication Standards Institute

EIRP Effective Isotropic Radiated Power

EIR Equipment Identity Register

E-UTRAN Evolved-UTRAN

EPC Evolved Packet Core

eNB evolved NodeB

EPS Evolved Packet System

XML Explainable Machine Learning

FCC Federal Communications Commission

FeMBB Further enhanced Mobile Broadband

FDM Frequency Division Multiplexing

FDMA Frequency Division Multiple Access

FDD Frequency Division Duplex

FD-MIMO Full-Dimension MIMO

FM Frequency Modulation

FSK Frequency Shift Keying

FHSS Frequency Hopping Spread Spectrum

1G First Generation

4G Fourth Generation

5G Fifth Generation

FFR Fractional Frequency Reuse

FBMC Filter Bank Multi Carrier

GSM Global System for Mobile Communications

GPRS General Packet Radio Service

GPS Global Positioning System

GMSK Gaussian Minimum Shift Keying

GoS Grade of Service

GGSN   Gateway GPRS Support Node

GMSK   Gaussian Minimum Shift Keying

GBR   Guaranteed Bit Rate

gNB   next generation NB

HD   High Definition

HMD   Head-Mounted Display

HTTP   Hypertext Transfer Protocol

HLR   Home Location Register

HSCSD   High Speed Circuit Switched Data

HSS   Home Subscriber Server

HCS   Human-Centric Services

HetVNets   Heterogeneous Vehicular Networks

ITU   International Telecommunication Union

ITU-T   International Telecommunication Union-Telecommunication Standards Sector

IoE   Internet of Everything

IoV   Internet of Vehicles

IoT   Internet of Things

IoST   Internet of Social Things

IIoT   Industrial Internet of Things

ICT   Information and Communications Technology

ISO   International Standards Organisation

IP   Internet Protocol

IEC   International Electrotechnical Commission

IETF   Internet Engineering Task Force

IEEE   Institute of Electrical and Electronics Engineers

ISM   Industrial, Scientific and Medical

IFA   Inverted-F Antennas

IFS	Inter-Frame Spacing
IRT	Intelligent Ray Tracing
ICIC	Inter-Cell Interference Coordination
IMSI	International Mobile Subscriber Identity
ICIC	Inter-Cell Interference Coordination
IBSS	Independent Basic Service Set
ITS	Intelligent Transportation Systems
KPI	Key Performance Indicator
LLC	Logical Link Control
LTE	Long-Term Evolution
LoS	Line-of-Sight
LA	Location Area
LAI	Location Area Identifier
LTE-A	LTE Advanced
LPWAN	Low Power Wide Area Network
LoRa	Long Range
MAC	Medium Access Control
MACA	Multiple Access with Collision Avoidance
MANO	Management and Orchestration Framework
ML	Machine Learning
MEC	Multi-access Edge Computing
mMTC	massive Machine Type Communications
M2M	Machine to Machine
Mulsemedia	MULtiple SEnsorial MEDIA
MIT	Massachusetts Institute of Technology
MOS	Mean Opinion Score
MSE	Mean Square Error
MTQI	Mobile TV Quality Index

MIMO Multiple Input Multiple Output

mMIMO massive MIMO

MU-MIMO Multi User-MIMO

mmWave millimeter Wave

MSK Minimum Shift Keying

MS Mobile Station

MADM Multi-Attribute Decision Making

MSC Mobile Switching Centre

MME Mobility Management Entity

MR Mixed Reality

MANO NFV Management and Orchestration

MBRLLC Mobile Broadband Reliable Low Latency Communication

mURLLC massive Ultra-Reliable and Low Latency Communications

MPS Multi-Purpose 3CLS and Energy Services

NLoS Non Line-of-Sight

NSS Network and Switching Subsystem

NSL Network Slicing

Non-GBR Non-Guaranteed Bit Rate

NR New Radio

NGC Next Generation Core

NGN Next Generation Networks

NFV Network Function Virtualization

NFVI Network Function Virtualization Infrastructure

NOMA Non Orthogonal Multiple Access

NAV Network Allocation Vector

NFC Near Field Communication

OSI Open Systems Interconnection

OFDM Orthogonal Frequency Division Multiplexing

OFDMA Orthogonal Frequency Division Multiple Access

OSS Operation and Support Subsystem

OVSF Orthogonal Variable Spreading Factor

OTT Over-The-Top

OpEx Operating Expenses

OBU On-Board Unit

OCB Outside the Context of a BSS

PDA Personal Digital Assistant

PC Personal Computer

PCF Point Coordination Function

PSNR Peak Signal to Noise Ratio

PEVQ Perceptual Evaluation of Video Quality

PSTN Public Switched Telephone Network

PSD Power Spectra Density

PIFA Planar Inverted-F Antennas

PM Phase Modulation

PSK Phase Shift Keying

PRMA Packet Reservation Multiple Access

PoA Point of Attachment

PIN Personal Identity Number

PMM Packet Mobility Management

PDN Packet Data Network

P-GW Packet Data Network Gateway

PAPR Peak to Average Power Ratio

PLATON Platform for Collaborative Learning and Teaching to Improve the Quality of Education through Disruptive Digital Technologies

PSM Power Save Mode

PSSCH Physical Sidelink Shared Channels

PSCCH Physical Sidelink Control Channels

QoE Quality of Experience

QoS Quality of Service

QPSK Quadrature Phase Shift Keying

QAM Quadrature Amplitude Modulation

QCI QoS Class Identifier

QoPE Quality of Physical Experience

RAT Radio Access Technology

RAN Radio Access Network

RL Reinforcement Learning

RF Radio Frequency

RRH Remote Radio Head

RTD Round Trip Delay

RTS Request to Send

RSS Relative Signal Strength

RA Routing Area

RAI Routing Area Identifier

RNC Radio Network Controller

RB Resource Block

RRM Radio Resource Management

RSU Road Side Unit

RFID Radio Frequency Identification

SMS Short Message Service

SAMMy Signal Strength-based Adaptive Multimedia Delivery Mechanism

SU-MIMO Single User-MIMO

SDM Space Division Multiplexing

SDMA Space Division Multiple Access

SNR Signal to Noise Ratio

2G   Second Generation

SIM   Subscriber Identity Module

SGSN   Serving GPRS Support Node

SC-FDMA Single-Carrier FDMA

SIR   Signal to Interference Ratio

S-GW   Serving Gateway

SDN   Software-Defined Networking

SIC   Successive Interference Cancellation

6G   Sixth Generation

TCP   Transmission Control Protocol

TDM   Time Division Multiplexing

TDMA   Time Division Multiple Access

TDD   Time Division Duplex

TACS   Total Access Communication System

3G   Third Generation

3GPP   Third Generation Partnership Project

TTI   Transmission Time Interval

UHD   Ultra High Definition

URLLC Ultra-Reliable and Low Latency Communications

umMTC ultra-massive Machine Type Communication

UMTS   Universal Mobile Telecommunications Service

UE   User Equipment

UTRAN UMTS Terrestrial Radio Access Network

URA   UTRAN Registration Area

VR   Virtual Reality

VoD   Video on Demand

VSQI   Video Streaming Quality Index

VTQI   Video Telephony Quality Index

VQA   Video Quality Assessment

VLC   Visible Light Communication

VLR   Visitor Location Register

VoLTE   Voice over LTE

VMNO   Virtual Mobile Network Operators

VNF   Virtual Network Function

VIM   Virtualized Infrastructure Manager

V2X   Vehicle-to-Everything

V2V   Vehicle-to-Vehicle

V2I   Vehicle-to-Infrastructure

V2N   Vehicle-to-Network

V2P   Vehicle-to-Pedestrian

VANET Vehicular Ad-Hoc Network

WCDMA Wideband CDMA

Wi-Fi   Wireless Fidelity

WLAN   Wireless Local Area Network

WRP   Wireless Routing Protocol

WAVE   Wireless Access in Vehicular Environments

WBSS   WAVE Basic Service Set

WSMP   WAVE Short Message Protocol

WPAN   Wireless Personal Area Network

WNAN wireless Neighborhood Area Network

WWAN Wireless Wide Area Network

ZRP   Zone Routing Protocol

Index

Printed in the United States
by Baker & Taylor Publisher Services